Springer Theses

Recognizing Outstanding Ph.D. Research

For further volumes:
http://www.springer.com/series/8790

Aims and Scope

The series "Springer Theses" brings together a selection of the very best Ph.D. theses from around the world and across the physical sciences. Nominated and endorsed by two recognized specialists, each published volume has been selected for its scientific excellence and the high impact of its contents for the pertinent field of research. For greater accessibility to non-specialists, the published versions include an extended introduction, as well as a foreword by the student's supervisor explaining the special relevance of the work for the field. As a whole, the series will provide a valuable resource both for newcomers to the research fields described, and for other scientists seeking detailed background information on special questions. Finally, it provides an accredited documentation of the valuable contributions made by today's younger generation of scientists.

Theses are accepted into the series by invited nomination only and must fulfill all of the following criteria

- They must be written in good English.
- The topic should fall within the confines of Chemistry, Physics, Earth Sciences, Engineering and related interdisciplinary fields such as Materials, Nanoscience, Chemical Engineering, Complex Systems and Biophysics.
- The work reported in the thesis must represent a significant scientific advance.
- If the thesis includes previously published material, permission to reproduce this must be gained from the respective copyright holder.
- They must have been examined and passed during the 12 months prior to nomination.
- Each thesis should include a foreword by the supervisor outlining the significance of its content.
- The theses should have a clearly defined structure including an introduction accessible to scientists not expert in that particular field.

Hai-Dong Wang

Theoretical and Experimental Studies on Non-Fourier Heat Conduction Based on Thermomass Theory

Doctoral Thesis accepted by
the Tsinghua University, Beijing, China

Author
Dr. Hai-Dong Wang
Department of Engineering Mechanics
School of Aerospace
Tsinghua University
Beijing
People's Republic of China

Supervisor
Prof. Zeng-Yuan Guo
Department of Engineering Mechanics
School of Aerospace
Tsinghua University
Beijing
People's Republic of China

ISSN 2190-5053 ISSN 2190-5061 (electronic)
ISBN 978-3-662-51337-8 ISBN 978-3-642-53977-0 (eBook)
DOI 10.1007/978-3-642-53977-0
Springer Heidelberg New York Dordrecht London

Softcover reprint of the hardcover 1st edition 2014

Printed on acid-free paper

Springer is part of Springer Science+Business Media (www.springer.com)

Parts of this thesis have been published in the following journal articles

SCI Papers:

1. Wang H D, Liu J H, Zhang X and Takahashi Koji, Breakdown of Wiedemann–Franz law in individual suspended polycrystalline gold nanofilms down to 3 K. International Journal of Heat and Mass Transfer, 2013, in press.
2. Wang H D, Liu J H, Guo Z Y, Zhang X, Zhang R F, Wei F and Li T Y, Thermal transport across the interface between a suspended single-wall carbon nanotube and air. Nanoscale and Microscale Thermophysical Engineering, 2013, 17: 349365.
3. Wang H D, Liu J H, Zhang X, Li T Y, Zhang R F and Wei F, Heat transfer between an individual carbon nanotube and gas environment in a wide Knudsen number regime. Journal of Nanomaterials, 2013, 181543.
4. Wang H D, Liu J H, Zhang X, Guo Z Y and Takahashi K. Non-Fourier heat conduction study for steady states in metallic nanofilms. Chinese Science Bulletin, 2012, 57(24): 3239−3243.
5. Wang H D, Cao B Y and Guo Z Y. Non-Fourier heat conduction in carbon nanotubes. Journal of Heat Transfer-ASME, 2012, 134, 051004.
6. Wang H D, Ma W G, Guo Z Y, Zhang X and Wang W. Measurement of electron–phonon coupling factor and interfacial thermal resistance of metallic nano-films using transient thermoreflectance technique. Chinese Physics B, 2011, 20(4), 040701.
7. Wang H D, Ma W G, Zhang X, Wang W and Guo Z Y. Theoretical and experimental study on the heat transport in metallic nanofilms heated by ultra-short pulsed laser. International Journal of Heat and Mass Transfer, 2011, 54(4): 967−974.
8. Wang H D, Liu J H, Zhang X, Guo Z Y and Takahashi K. Experimental study on the influences of grain boundary scattering on the charge and heat transport in gold and platinum nanofilms. Heat and Mass Transfer, 2011, 47(8): 893−898.
9. Wang H D and Guo Z Y. Thermon gas as the thermal energy carrier in gas and metals. Chinese Science Bulletin, 2010, 55(29): 3350−3355.
10. Wang H D, Cao B Y and Guo Z Y. Heat flow choking in carbon nanotubes. International Journal of Heat and Mass Transfer, 2010, 53(9−10): 1796−1800.
11. Wang H D, Ma W G, Zhang X and Wang W. Measurement of the thermal wave in metal films using femtosecond laser thermoreflectance system (in Chinese). ACTA PHYS SIN-CH ED, 2010, 59(6): 3856−3862.
12. Zhang R F, Zhang Y Y, Zhang Q, Xie H H, Wang H D, Nie J Q, Wen Q and Wei F. Optical visualization of individual ultralong carbon nanotubes by chemical vapour deposition of titanium dioxide nanoparticles. Nature Communications, 2013, 4: 1727.

13. Liu J H, Wang H D, Ma W G, Zhang X and Song Y. Simultaneous measurement of thermal conductivity and thermal contact resistance of individual carbon fibers using Raman spectroscopy. Review of Scientific Instruments, 2013, 84: 044901.
14. Ma W G, Wang H D, Zhang X and Wang W. Theoretical and experimental study of femtosecond pulse laser heating on thin metal film (in Chinese). ACTA PHYS SIN-CH ED, 2011, 60(6): 064401.
15. Ma W G, Wang H D, Zhang X and Wang W. Study of the electron–phonon relaxation in thin metal films using transient thermoreflectance technique. International Journal of Thermophysics, 2011, DOI: 10.1007/s10765-011-1063-2.
16. Ma W G, Wang H D, Zhang X and Wang W. Experiment study of the size effects on electron–phonon relaxation and electrical resistivity of polycrystalline thin gold films. Journal of Applied Physics, 2010, 108: 064308.
17. Ma W G, Wang H D, Zhang X and Takahashi Koji. Different effects of grain boundary scattering on charge and heat transport in polycrystalline platinum and gold nanofilms. Chinese Physics B, 2009, 18(5): 2035–2040.

EI Papers:

1. Wang H D and Guo Z Y. Thermomass two step model and general heat conduction law for metals (in Chinese). Journal of Engineering Thermophysics, 2013, 34(4): 738–741.
2. Wang H D, Ma W G, Guo Z Y, Zhang X and Wang W. Experimental study of ultra-fast heat conduction process in metals using femtosecond laser thermal reflection method (in Chinese). Journal of Engineering Thermophysics, 2011, 32(3): 465–468.
3. Wang H D, Cao B Y and Guo Z Y, Motion of thermomass in metals—state equation for thermomass in electron gas (in Chinese). Journal of Engineering Thermophysics, 2010, 31(5): 817–820.
4. Liu J H, Wang H D, Ma W G, Zhang X and Guo Z Y, Experimental study of thermal and electrical properties of gold nanofilms at ultra low temperature (in Chinese). Journal of Engineering Thermophysics, 2012, 33(11): 1944–1946.
5. Ma W G, Wang H D, Zhang X and Wang W. Electron–phonon coupling in thin gold films (in Chinese). CIESC Journal, 2011, 62(S1): 48–53.
6. Ma W G, Wang H D, Zhang X and Guo Z Y. Experimental study of electron–phonon coupling factor of copper thin film (in Chinese). Journal of Engineering Thermophysics, 2010, 31(3): 499–502.
7. Ma W G, Wang H D, Cao B Y and Zhang X. Experimental study of thermal and electrical properties of gold nanofilms (in Chinese). Journal of Engineering Thermophysics, 2009, 30(11): 1907–1909.

International Conference Papers:

1. Wang H D, Cao B Y, Zhang X and Guo Z Y. Experimental proof of energy-mass duality of heat. In: Proceedings of 3rd International Forum on Heat Transfer (IFHT2012). November 13–15, 2012, Nagasaki, Japan.
2. Wang H D, Ma W G, Zhang X and Wang W. Use of genetic algorithms for the simultaneous estimation of electron–phonon coupling factor and interfacial thermal resistance of metallic thin films. In: Proceedings of 9th Asian Thermophysical Properties Conference (ATPC 2010). October 19–22, 2010, Beijing, China.
3. Wang H D, Ma W G, Zhang X and Wang W. Mass nature of heat and its applications iv: thermal wave and periodic temperature oscillation in metallic films heated by ultra-short pulsed lasers. In: Proceedings of 14th International Heat Transfer Conference (IHTC 2010). August 8–13, 2010, Washington DC, America.
4. Wang H D, Ma W G, Zhang X and Guo Z Y. Measurement of the thermal wave in metal films using femtosecond laser thermoreflectance system. In: Proceedings of 9th Kyoto-Seoul National-Tsinghua University Thermal Engineering Conference. October 21–23, 2009, Kyoto, Japan.
5. Wang H D, Cao B Y and Guo Z Y. Non-Fourier heat conduction in carbon nanotubes. In: Proceedings of 2nd Micro/Nanoscale Heat & Mass Transfer International Conference (MNHMT 2009). December 18–21, 2009, Shanghai, China.
6. Ma W G, Wang H D, Zhang X, et al. A novel relationship between thermal and electrical conductivities in polycrystalline metallic nanofilms. In: Proceedings of the 20th International Symposium on Transport Phenomena. July 7–10, 2009, Victoria B C., Canada.

Supervisor's Foreword

Recent rapid developments of ultra-fast laser technique and ultra-high heat flux micro-processors have imposed great challenges to classical thermophysical sciences. The fundamental theory of heat conduction Fourier's law is no longer valid for these extreme conditions. In recent years, great efforts have been made to study non-Fourier heat conduction, but the experimental data available are still limited because the non-Fourier phenomena only occur at very low temperatures or ultra-short time scales. Also, the present non-Fourier heat conduction models are only phenomenological ones depending on empirical parameters. Lacking a thorough understanding of the macroscale physical mechanisms, microscopic theory is difficult to use for practical applications. No theoretical models have yet been developed that fully explain non-Fourier heat conduction. This thesis analyzes non-Fourier heat conduction based on the first principles to develop a general heat conduction law. The theory is validated by comparisons with experimental results.

The main content and conclusions are:

1. Thermomass theory is used to analyze non-Fourier heat conduction. Thermomass is the relativistic mass of heat and the heat flux is known as the directional flow of thermomass along a temperature gradient. Newtonian mechanics is used to establish the thermomass motion equation, which is actually the general heat conduction equation. The thermomass theory gives a full understanding of non-Fourier heat conduction as the consequence of the non-negligible thermomass inertia effect. Furthermore, thermomass theory predicts the occurrence of non-Fourier heat conduction even for the steady case for the first time.
2. A femtosecond laser thermoreflectance system has been established to detect the ultra-fast heat transfer between electrons and phonons in metallic nanofilms. A temperature wave was observed with a propagation speed of about 8.1×10^5 ms^{-1}. The temperature wave is distinguished from the thermal wave with the temperature wave being the heat diffusion for periodic boundary conditions, while the thermal wave is actually the hyperbolic wave propagation.
3. A low temperature direct current measurement system has been established to study steady state non-Fourier heat conduction. The measured average temperature of the gold nanofilm was notably higher than the temperature predicted by Fourier's law, with the temperature difference increasing as the heating power increased or the environmental temperature decreased. The maximum

temperature difference reached 23 K at an environmental temperature of 3 K when the heat flux exceeded 2×10^{10} Wm^{-2}. In this case, the thermomass inertia is non-negligible which causes the deviation from Fourier's law. The good agreement between the predictions of the general heat conduction law and the experimental data validates the thermomass theory.

4. The electrical and thermal conductivities of several gold nanofilms were measured from 3 K to 300 K. The measured conductivities are much less than the corresponding bulk conductivities, showing a significant size effect. The electron-grain boundary scattering is shown to be the dominant factor for this size effect. The Wiedemann–Franz law is found to fail at low temperatures because of the inelastic electron scattering (Raman electron scattering). A new theoretical model that takes inelastic electron scattering into account agrees well with the experimental data.

Beijing, October 2013 Zeng-Yuan Guo

Acknowledgments

I would like to express my appreciation and thanks to my supervisor, Prof. Zeng-Yuan Guo, and to Prof. Xing Zhang who has guided and helped me with my experimental work from the very beginning. It has been a great honor for me to be a student of these two great professors in the past several years. They have guided me into this exciting frontier thermal science field and taught me how to think and work as a skilled researcher. I appreciate all their ideas, encouragement, and constructive comments that help me to accomplish my Ph.D. research at Tsinghua University.

I would also like to thank all the members in our research group, Wei-Gang Ma, Jin-Hui Liu, Qin-Yi Li, Yu-Dong Hu, Jian-Li Wang, Xue-Tao Cheng, Wei-Ming Song, Zhi-Qiang Zhou, Sheng-Hong Ju, and Yuan-Wei Li. Their brilliant insights and collaboration have been of great help to me. Their warmness and kind sharing made our laboratory a great family.

I would like to thank Prof. Koji Takahashi, Prof. Hiroshi Takamatsu and Prof. Yasuyuki Takata at Kyushu University for their excellent nanofilm samples and kind suggestions for experiments and analyses. I would like to thank Prof. Kai-Li Jiang, Prof. Fei Wei, and their students, Tian-Yi Li, Ru-Fan Zhang and Huan-Huan Xie for providing us perfect carbon nanotube samples and kind discussions for experiments.

Finally, I would like to thank my wife and parents for their never-ending love and support for my research career.

This work is supported by the National Natural Science Foundation of China (Grant Nos. 51327001, 51136001, 51076080, 50730006 and 50976053), China Postdoctoral Science Foundation and Tsinghua University Initiative Scientific Research Program.

Contents

Chapter 1
Introduction

Abstract With the rapid development of femtosecond laser heating, integrated circuit, micro/nano electromechanical systems, the research on heat transfer in nanomaterials under high heat flux conditions has attracted increased attention. The traditional Fourier's law is found to be broken under these extreme conditions. It is in urgent need to develop a general heat conduction model to replace Fourier's law and give precise predictions for thermal analysis in practical applications. This thesis reports on the theoretical and experimental studies of non-Fourier heat conduction using a novel thermomass theory as basis. A femtosecond laser thermoreflectance system and a direct current electrical measurement system at liquid helium temperature have been established for experimental investigations. The heat transfer behaviors under the extreme conditions have been studied in-depth and the experimental data were utilized to verify the theoretical models. This chapter introduces the background of non-Fourier heat conduction and the recent research on the material properties of metallic nanofilms.

1.1 Present Study on Non-Fourier Heat Conduction

In 1804, French physicist Boit analyzed the experimental data of steady heat conduction between two parallel plates and gave the empirical relation [1]:

$$Q = \kappa \frac{\Delta T}{\delta} A \tag{1.1}$$

where Q is the heat transferred per unit time, A and δ are the surface area and thickness of the plate, ΔT is the temperature difference, and κ is the thermal conductivity. Later in 1822, another French physicist and mathematician Fourier investigated the experimental results of heat conduction in-depth and gave a more general expression:

$$q = -\kappa \nabla T \tag{1.2}$$

H.-D. Wang, *Theoretical and Experimental Studies on Non-Fourier Heat Conduction Based on Thermomass Theory*, Springer Theses, DOI: 10.1007/978-3-642-53977-0_1,

where q is the heat flux. This is the famous Fourier's law, which becomes the theoretical foundation of heat transfer. Although Fourier's law is only an empirical relation, it captures the general relationship between heat flux and temperature gradient, i.e., a constant proportional coefficient exists. This coefficient is referred to as thermal conductivity, which is seen as one of the material properties. To become a material property, the thermal conductivity should be independent of the geometrical dimensions and heat flux, decided only by the material basic characteristics, such as its molecular composition, grain size, etc.

Fourier's law demonstrates that the physical essence of heat conduction is diffusion. Yet this causes a paradox that the propagation speed of thermal disturbance is infinite, which disagrees with some experimental observations. Peshkov [2] found in the experiment that heat was transferred as a form of wave in superfluid helium with a certain propagation speed. This phenomenon was called "second sound" and broke Fourier's law. The first sound was the mechanical wave and the second sound was the thermal wave. Later, Landau [3] proposed a two-fluid model to describe the thermal wave behaviors at ultra-low temperatures, the liquid helium could be seen as a mixture of normal fluid and superfluid below 2.17 K. Then Ward [4] explained the thermal wave in liquid helium based on a phonon theory. Chester [5] found second sound phenomenon in solid crystals. Narayanamurti [6] measured the speed of second sound in bismuth. Brorson [7] measured the propagation speed of femtosecond laser pulses in metallic films and found it close to Fermi velocity $10^6\ \mathrm{ms}^{-1}$ at room temperature.

For theoretical study of thermal waves, Cattaneo [8], Vernotte [9], Morse and Feshbach [10] modified the Fourier's equation by introducing a relaxation time. In this way, the heat diffusion equation was transferred to a hyperbolic wave equation, which was referred to as C-V model. Barletta and Zanchini [11] calculated the entropy generation based on C-V model and found negative value for semi-infinite body under the time-dependent heat flux boundary conditions; this is against the second law of thermodynamics. In this case, the assumption of local thermal equilibrium is not valid and the C-V model needs to be modified. Lor and Chu [12] calculated the thermal wave propagation in the film deposited on substrate using a numerical method. The result showed that part of the thermal wave could be reflected at the interface, which was totally different from the prediction of heat diffusion model. Some analytical solutions of the thermal wave equation were obtained in Refs. [13–19] for different laser source functions and boundary conditions, giving wave-like temperature responses. Reference [20] pointed out that when the frequency of varied boundary temperature was comparable with the reciprocal of the relaxation time, the thermal wave effect became non-negligible compared with the heat diffusion. Meanwhile, the effect of nonlinear thermal conductivity on the thermal wave propagation process was analyzed in Refs. [21, 22].

Besides the C-V model, Tzou [23] developed a dual-phase lag (DPL) model by introducing two relaxation times responsible for heat flux and temperature gradient. Dai and Su [24, 25] compared the C-V model and DPL model and found that C-V model was the approximation result of DPL model for small relaxation times. But it should be noted that all these thermal wave models were phenomenological models depending on the assumption of relaxation times, though the understanding of the physical essence of thermal wave phenomenon is still lacking.

Besides the transient thermal wave phenomenon, the ballistic transport of heat in steady states can also result in non-Fourier heat conduction behaviors. Based on Boltzmann transport equation, Majumdar [26] derived a governing equation for heat transfer by phonons, which was capable of describing the phonon wave phenomenon. This equation could be reduced to Fourier's equation under the macro-limit conditions and transformed to black body radiation equation under the micro-limit conditions. Chen [27] gave a ballistic-diffusion transport equation considering these two effects for phonons. When the ballistic transport rules, the mean free path of phonons is dependent on the system scales, thus the thermal conductivity becomes size-dependent and cannot be seen as a simple material property. This is against the linear relationship between heat flux and temperature gradient described by Fourier's law. In this case, some other process parameters should be taken into consideration.

Based on the present studies of non-Fourier heat conduction, one can conclude that: (1) the existing modified models of Fourier's law are far from perfect, the physical mechanism of non-Fourier phenomenon should be investigated in-depth; (2) the experimental studies of non-Fourier heat conduction are rare and the theoretical investigation should be based on the convincing experimental data. Thus this thesis gives a systematic study of non-Fourier heat conduction using theoretical and experimental methods. For theoretical study, a general heat conduction law has been developed based on a novel thermomass theory, which reveals the inner relationship between thermal science and other scientific fields, such as mechanics and electricity. For experimental study, a femtosecond laser thermoreflectance system and a direct current electrical measurement system have been established to observe the non-Fourier heat conduction behaviors in unsteady and steady states.

1.2 Present Theoretical Models

Fourier's law will result in a paradox of infinite propagation speed of thermal disturbance. Although this paradox can be neglected at normal timescales, Fourier's law should be modified at femtosecond scales. Some popular modified models are listed here.

1.2.1 C-V Model

In 1958, Cattaneo [8] and Vernotte [9] developed a modified model of Fourier's law by introducing a relaxation time of heat flux. The heat diffusion equation can be changed into a hyperbolic wave equation. The one-dimensional C-V model with a internal heat source is shown as

$$\frac{\partial q}{\partial x} + \rho C \frac{\partial T}{\partial t} = S \tag{1.3}$$

$$\tau_0 \frac{\partial q}{\partial t} + q = -\kappa \frac{\partial T}{\partial x} \tag{1.4}$$

where S and τ_0 are the internal heat source per unit volume and relaxation time, respectively. κ is the thermal conductivity. C-V model attributes the thermal wave to the lagging effect between heat flux and temperature gradient. But the accurate expression for the relaxation time is missing, C-V model is only a phenomenological model. A temperature governing equation can be derived from Eqs. 1.3 and 1.4 as

$$\tau_0 \frac{\partial^2 T}{\partial t^2} + \frac{\partial T}{\partial t} = \frac{\kappa}{\rho C} \frac{\partial^2 T}{\partial x^2} + \frac{1}{\rho C} \left(S + \tau_0 \frac{\partial S}{\partial t} \right) \tag{1.5}$$

It is seen that Eq. 1.5 is a damped wave equation, which will reduce to Fourier's heat diffusion equation when $\tau_0 = 0$. The sound speed is given as

$$C_h = \sqrt{\frac{\kappa}{\rho C \tau_0}} \tag{1.6}$$

1.2.2 Hyperbolic Two-Step Model

Anisimov [28] studied the heat conduction process in metallic films heated by ultra-short pulsed laser and developed a two-temperature model including electron temperature and lattice temperature. Based on it, Qiu and Tien [29–31] developed a hyperbolic two-step (HTS) model in 1993 by adding a heat flux relaxation time. One-dimensional HTS model can be expressed as

$$\rho_e C_e \frac{\partial T_e}{\partial t} = -\frac{\partial q_e}{\partial x} - G(T_e - T_l) + S \tag{1.7}$$

$$\tau_e \frac{\partial q_e}{\partial t} + q_e = -\kappa_e \frac{\partial T_e}{\partial x} \tag{1.8}$$

$$\rho_l C_l \frac{\partial T_l}{\partial t} = -\frac{\partial q_l}{\partial x} + G(T_e - T_l) \tag{1.9}$$

$$\tau_l \frac{\partial q_l}{\partial t} + q_l = -\kappa_l \frac{\partial T_l}{\partial x} \tag{1.10}$$

where the subscripts e and l stand for electron and lattice; τ_e and τ_l are the relaxation times for electrons and lattices, respectively. G is the electron–phonon coupling factor. Equations 1.8 and 1.10 have the same expression as C-V model. For metals, τ_e is much smaller than τ_l, thus the electrons absorb the energy of photons in the first step and the electron temperature T_e increases rapidly. After that, the electrons transport their energy to the lattices by colliding with phonons, the lattice temperature

T_l increases while T_e decreases. The governing equation of electron temperature can be obtained based on Eqs. 1.7 and 1.8 as

$$\tau_e \rho_e C_e \frac{\partial^2 T_e}{\partial t^2} + (\rho_e C_e + \tau_e G) \frac{\partial T_e}{\partial t} = \kappa_e \frac{\partial^2 T_e}{\partial x^2} + \tau_e G \frac{\partial T_l}{\partial t} + \tau_e \frac{\partial S}{\partial t} - G(T_e - T_l) + S \tag{1.11}$$

Equation 1.11 is a damped wave equation, the propagation speed of thermal waves in the electron system is

$$C_h = \sqrt{\frac{\kappa_e}{\rho_e C_e \tau_e}} \tag{1.12}$$

1.2.3 Parabolic Two-Step Model

When the laser pulse is much longer than the relaxation time of electrons, HTS model will reduce to the parabolic two-step (PTS) model. One-dimensional PTS model can be expressed as

$$\rho_e C_e \frac{\partial T_e}{\partial t} = \kappa_e \frac{\partial^2 T_e}{\partial x^2} - G(T_e - T_l) + S \tag{1.13}$$

$$\rho_l C_l \frac{\partial T_l}{\partial t} = G(T_e - T_l) \tag{1.14}$$

where Eq. 1.13 is a heat diffusion equation for zero relaxation time. The governing equation of electron temperature can be obtained as

$$\rho_e C_e \frac{\partial^2 T_e}{\partial t^2} + G\left(1 + \frac{\rho_e C_e}{\rho_l C_l}\right) \frac{\partial T_e}{\partial t} = \frac{\kappa_e G}{\rho_l C_l} \frac{\partial^2 T_e}{\partial x^2} + \kappa_e \frac{\partial^3 T_e}{\partial t \partial x^2} + \frac{\partial S}{\partial t} + \frac{GS}{\rho_l C_l} \tag{1.15}$$

where the first term on the left-hand side of equal sign is much smaller than the second term on the left-hand side; thus Eq. 1.15 can be simplified as

$$G\left(1 + \frac{\rho_e C_e}{\rho_l C_l}\right) \frac{\partial T_e}{\partial t} = \frac{\kappa_e G}{\rho_l C_l} \frac{\partial^2 T_e}{\partial x^2} + \kappa_e \frac{\partial^3 T_e}{\partial t \partial x^2} + \frac{\partial S}{\partial t} + \frac{GS}{\rho_l C_l} \tag{1.16}$$

Equation 1.16 is a typical parabolic equation. Most of the energy of metals is contained by lattices, while the electrons are the main energy carriers. The coupling factor G appears to be the "bridge" between electrons and lattices, describing the strength of electron–phonon interactions.

1.2.4 Phonon Kinetic Model

Based on the linearized Boltzmann transport equation, Guyer and Krumhansl [32, 33] proposed a phonon hydrodynamics model to describe the thermal wave propagation in dielectrics. In this model, a set of macroscopic equations was derived to describe the Poiseuille flow in a phonon gas, based on which, the second sound behavior could be investigated. Meanwhile, the normal and umklapp processes of phonon interactions were considered in the model. A typical phonon hydrodynamics model can be expressed as

$$C_p \frac{\partial T}{\partial t} + \nabla q = 0 \tag{1.17}$$

$$\frac{\partial q}{\partial t} + \frac{c^2 C_p}{3} \nabla T + \frac{1}{\tau_R} q = \frac{\tau_N c^2}{5} \left[\nabla^2 q + 2\nabla \left(\nabla q \right) \right] \tag{1.18}$$

where τ_R and τ_N are the relaxation times of umklapp and normal processes, respectively. Further, the governing equation of temperature can be obtained as

$$\nabla^2 T + \frac{9\tau_N}{5} \frac{\partial}{\partial t} \left(\nabla^2 T \right) = \frac{3}{\tau_R c^2} \frac{\partial T}{\partial t} + \frac{3}{c^2} \frac{\partial^2 T}{\partial t^2} \tag{1.19}$$

It is seen that Eq. 1.19 is a damped hyperbolic wave equation; several phonon scattering mechanisms can be studied in detail. The speed of thermal waves in dielectrics is derived as

$$C_h = \frac{C_s}{\sqrt{3}} \tag{1.20}$$

where C_s is the sound speed.

1.2.5 Dual-Phase Lag Model

Tzou [23] developed a dual-phase lag (DPL) model considering different relaxation times for heat flux and temperature gradient. The heat conduction behaviors from diffusion to thermal waves can be described using the same model. A complete expression for DPL model is

$$q(x, t + \tau_q) = -\kappa \frac{\partial T(x, t + \tau_T)}{\partial x} \tag{1.21}$$

$$-\frac{\partial q}{\partial x} + S = \rho C \frac{\partial T}{\partial t} \tag{1.22}$$

where τ_q and τ_T are the relaxation times for heat flux and temperature gradient, respectively. Using Taylor expansion and ignoring the second-order small quantities, one-dimensional governing equation of temperature can be obtained as

$$\frac{\partial^2 T}{\partial x^2} + \tau_T \frac{\partial^3 T}{\partial x^2 \partial t} + \frac{1}{\kappa} \left(S + \tau_q \frac{\partial S}{\partial t} \right) = \frac{1}{\alpha} \frac{\partial T}{\partial t} + \frac{\tau_q}{\alpha} \frac{\partial^2 T}{\partial t^2} \tag{1.23}$$

By choosing $\alpha = \frac{\kappa}{\rho_e C_e + \rho_l C_l}$, $\tau_T = \frac{\rho_l C_l}{G}$ and $\tau_q = \tau_e + \frac{1}{G}\left(\frac{1}{\rho_e C_e} + \frac{1}{\rho_l C_l}\right)^{-1}$, the DPL model will be transformed into the HTS model.

1.3 Present Experimental Study of Heat Conduction in Metallic Nanofilms

In order to study the non-Fourier heat conduction behaviors experimentally, we need to build different experimental systems for unsteady and steady states measurements. In unsteady states, a femtosecond laser pump-probe technique can be used to ensure a high time resolution at femtosecond scales. In steady states, a large current electrical measurement system can be used to study the steady non-Fourier behavior under the high heat flux, low temperature conditions. In this chapter, we begin with a brief review of the experimental research history and then come to the goal of our experimental research.

1.3.1 Experimental Study in Unsteady States

With the rapid development of the picosecond and femtosecond laser techniques, the ultra-short pulsed laser has been widely used to investigate the very fast physical processes. A pump-probe technique has been developed to achieve a high time resolution at femtoseconds [34]. The electron relaxation time in metals is about 10^{-14} s, making the pump-probe technique the only possible way. Some related research results are listed in Table 1.1:

The timescale of experimental investigation has been widely extended by using the ultra-short pulsed lasers, making it possible to observe the rapid interactions of electrons and phonons in metals. All the experiments in Table 1.1 were performed using a transient thermoreflectance (TTR) method. In this method, a femtosecond laser was split into a pump beam and a probe beam. The pump beam was used to heat the sample and the probe beam was used to monitor the change of thermoreflectance, which was proportional to the change in temperature. By precisely controlling the difference between the light paths of these two beams, a femtosecond time resolution could be achieved. The details of the experiment will be discussed in the Chap. 3.

The femtosecond laser pump-probe technique has been discussed in detail in Ref. [45]. Figure 1.1 gives the schematic diagram of the pump-probe system. Using this technique, some material properties have been measured and listed in Table 1.2: For theoretical study of the heat conduction caused by femtosecond laser heating, some numerical simulations of thermal wave propagation based on the DPL model have been reported in Refs. [61–64]. The temperature dependence of electron specific heat and thermal conductivity has been studied using a molecular dynamics method in Ref. [65], the non-equilibrium energy exchange between the electrons and lattices

Table 1.1 Experimental results of the metals heated by ultra-fast lasers

Time/Researcher	Method	Purpose	Main conclusions
1984/Eesley [35, 36]	Transient thermoreflectance (TTR) method	Testing Metallic nanofilms	(1) A non-equilibrium heat conduction process exists between electrons and lattices;
			(2) Obtain the heat diffusivity of metallic films by analyzing the transient temperature response;
			(3) Prove the validation of TTR method
1984/Fujimoto [37]	TTR method	Testing tungsten samples	A 75 fs laser was used for measurement, the obtained temperature response reflects the non-equilibrium heat conduction process
1987/Elsayed-Ali [38, 39]	TTR method	Testing copper films	A femtosecond laser (150 fs–300 fs) was used for measurement, the non-equilibrium energy exchange between electrons and lattices has been observed
1987/Brorson [7]	TTR method	Observing temperature waves in metals	(1) A rear heating-front detecting scheme was used for measurement;
			(2) The measured speed of temperature wave was close to the Fermi speed
1990/Brorson [29]	TTR method	Measuring coupling factors	(1) The measured results agreed well with Allens two-temperature model [40];
			(2) The characteristic time of electron–phonon coupling was found to be several picoseconds
1991/Elsayed-Ali [41]	TTR method	Testing monocrystal and polycrystal films	(1) It was confirmed that the characteristic coupling time increases as the heating power increases;
			(2) When the Au film thickness is close the optical penetration thickness, the coupling time in monocrystal is shorter
1992/Juhasz [42]	TTR method	Testing the influence of ambient temperature	(1) Define a an effective electron temperature of single-crystal thin Au films;
			(2) Observe the influence of the ambient temperature on the decay of the effective electron temperature

(continued)

Table 1.1 (continued)

Time/Researcher	Method	Purpose	Main conclusions
1994/Tien and Qiu [29, 30]	TTR method	Testing Au/Cr nanofilms	(1) The experimental data agreed well with the prediction of the PTS model; (2) Prove the feasibility of using numerical simulations to analyze the ultra-fast heat conduction process
1994/Sun [43]	TTR method	Observing thermalization process of electrons	(1) Observe the non-equilibrium heat conduction process between electrons and phonons; (2) The experimental data agreed well with the prediction of Boltzmann equation, the thermal-ization time of electrons was estimated to be about 1.5 ps
1997/Hohlfeld [44]	TTR method	Testing Au nanofilms	(1) A front heating-front detecting scheme was used for measurement; (2) The mean free path of electrons was about 100 nm at 1500 K

plays a dominant role in the first several picoseconds. Based on the two-temperature model, Ref. [66] studied the electron–phonon coupling factor in different metals, such as: Au, Ag, Cu, and Al. Ref. [67] developed a multi-material two-temperature model for the simulation of ultra-short pulsed laser ablation. Reference [68–70] developed a finite difference method to solve the PTS model numerically, which was suitable for single-layer and double-layer thin films with interfacial thermal resistance. Saidane and Pulko [71] studied the high-power ultra-short pulsed laser heating of low-dimensional structures using a transmission line method (TLM), which was efficient for analysis of multi-layered thin films. Reference [72] calculated the temperature and thermal stress responses induced by femtosecond laser heating. Reference [73] studied the non-Fourier heat conduction in silicon using a lattice Boltzmann method. The propagation speed of thermal wave could equal the speed of sound at the ballistic transport limit, some related research reports could be found in Refs. [74, 75].

In TTR system, a femtosecond laser beam is split into a pump beam and a probe beam, where the pump beam is used to heat the film sample and the probe beam is used to detect the change of the reflectivity of the film surface. A precise stepping motor is used to control the optical path delay between these two beams and a femtosecond time resolution can be obtained. An acousto-optic modulator (AOM) is used to create a train of modulated pump pulses. The focused laser spots of two beams are made to coincide with each other according to the images from a CCD camera.

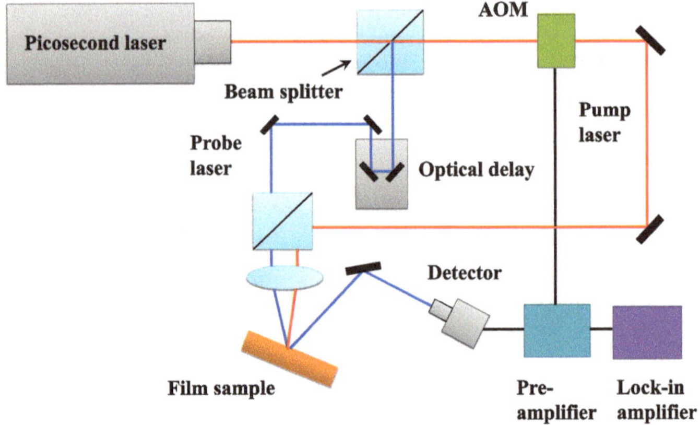

Fig. 1.1 Schematic diagram of the pump-probe system [45]

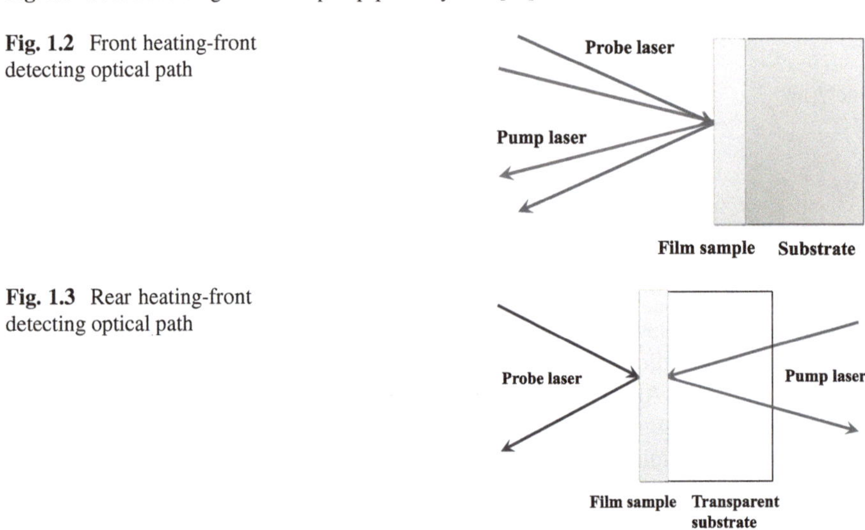

Fig. 1.2 Front heating-front detecting optical path

Fig. 1.3 Rear heating-front detecting optical path

Figures 1.2 and 1.3 show two kinds of pump-probe optical paths, i.e., front heating-front detecting and rear heating-front detecting, respectively. In Fig. 1.2, the pump and probe beams are both focused at the front surface of the film sample, which is suitable for the film deposited on the opaque substrate. In Fig. 1.3, the pump and probe beams are focused separately on both sides of the film sample, which is suitable for the transparent substrate. In this way, the scattered pump laser is blocked by the film sample and the signal-to-noise ratio can be increased.

In this thesis, an experimental system similar to Fig. 1.1 has been established and aims to: (1) study the transient heat conduction process induced by femtosecond laser heating and measure the speed of temperature wave; (2) measure the transient material properties of metallic nanofilms, such as electron–phonon coupling factor, interfacial thermal resistance, and thermal conductivity of substrate.

Table 1.2 Experimental results of material properties using TTR method

Researcher	Material or purpose	Main conclusions
Brorson et al. [46]	Solid C_{60} film	(1) Observe the non-linear optical response of the solid C_{60} film by analyzing the measured thermoreflectance signals; (2) The electron relaxation time of the solid C_{60} film was measured to be less than 1 ps; (3) The electron-phonon coupling factor reflects the strength of interactions between electrons and phonons [47, 48]
Norris et al. [49–56]	To measure the electron-phonon coupling factor and interfacial thermal resistance	The electron-phonon coupling factor was measured to be dependent to the substrate materials and metallic films thickness
Capinski et al. [57]	To measure the thermal conductivity of GaAs/AlAs superlattice films	The measured values were remarkably lower than the bulk value, the phonon–phonon and phonon-defect scattering could be the main reason
Nakamiya et al. [58]	Multi-walled carbon nanotube films	The maximum electron temperature was measured to be 1400 °C through the shifted D-band and G-band frequencies, consistent with the TTR measurement results
Woutersen et al. [59]	Water molecules with different hydrogen-bonds	Combine the TTR method with infrared spectrum analysis and confirm the existence of two distinct molecular species in water by observe the orientational relaxation of water molecules at different time scales
Tokmakoff et al. [60]	Polyatomic liquids	Investigate the temperature dependencies of the thermal populations of phonons, the phonon density of states and the anharmonic coupling matrix elements using TTR method

1.3.2 Experimental Study in Steady States

With the rapid development of micro/nano electro-mechanical systems (MEMS/NEMS) technique, the thermal management in micro/nano scales becomes a hot research topic [76, 77]. As the interconnection in micro/nano electronic devices, the metallic nanofilms have attracted much attention. The charge and heat transport in metallic nanofilms has been studied theoretically and experimentally, and some important achievements are listed as follows.

1. Theoretical Study

In earlier times, Thomson [78] and Lovell [79] have found that the electrical conductivity of metallic thin film was smaller than the corresponding bulk value, which was referred to as the size effect. Then, Fuchs [80] and Sondheimer [81] studied the size effect based on the free electron gas theory and Boltzmann transport equation, and developed an FS model considering electron scattering at the film surfaces. But the FS model ignored the electron scattering at the grain boundaries, making it unacceptable for the polycrystalline nanofilms. To solve this problem, Mayadas and Shatzkes [82, 83] modified the FS model and introduced a reflection factor to estimate the effect of electron scattering at grain boundaries. The FS model and MS model became the theoretical foundation for the electrical and thermal conductivities of the metallic nanofilms.

In practical applications, an electrical–thermal analogy has been used to predict the thermal conductivity based on the easily measured electrical conductivity. This analogy was given by the Wiedemann-Franz law, which stated the fact that the electrons transported charge and heat by a constant ratio. Tien [84, 85] developed a prediction model of the thermal conductivity combining the FS model and the Wiedemann-Franz law. Then Qiu [29] and Kumar [19] developed a more accurate model combining the MS model, Boltzmann transport equation, and Wiedemann-Franz law, where the different electron scattering effects were considered using the Matthiessen's rule [85]. Given the proper grain size and grain boundary reflection factor, this model matched well with the experimental data.

2. Experimental Study

Many measurement methods have been developed for specific materials [86, 87], such as infrared thermography [88], Raman spectroscopy measurement [89], optical thermal sensor [90], optical acoustic coupling method [91], photodiodes [92], near-field optical temperature measurement [93], scanning thermal microscopy [94], fluorescence method [95], thermoelectric couple and resistance thermometer [96], etc. But only a few methods can be applied for the materials at nanoscales due to three main reasons: (1) Low spatial resolution. The temperature sensors of the thermoelectric couple or resistance thermometer is even bigger than the tested sample; (2) Low temperature resolution. Many optical or electrical measurement methods need calibration, but the temperature of nanomaterial is sensitive to the environment, the calibration process will bring uncertainty; and (3) Contact methods will bring extra thermal resistance and uncertainty for measurements.

For metallic nanofilms, direct current heating method, 3ω method and optical thermal sensor method are normally used in the experiments. Some typical experimental results are listed as follows. For monocrystalline films, Duggal [97, 98], Kirkland [99], Caballero [100], and Kästle [101] measured the electrical conductivities of Ag, Al, Ti, and Au nanofilms, and found the FS model valid when the film thickness was larger than the electron mean free path (MFP) at high temperatures. Otherwise, the nanostructure inside the metallic film sample would affect the macro material properties.

For polycrystalline films, Ramaswamy [102], Fenn [103], Durkan [104], Wu [105], and Maroma [106] measured the electrical conductivities and temperature coefficients of resistance (TCR) of Cu, Nb, and Au nanofilms, and found that the electron scattering at film surfaces and grain boundaries would affect the electrical conductivity significantly. The surface scattering played a dominant role when the electron MFP was much larger than the grain size, while the grain boundary scattering became dominant when the electron MFP was comparable with the grain size or even smaller. The MS model should be used for the latter case. Because the electron MFP increased as the temperature decreased, the electron grain boundary scattering became more important at low temperatures. Meanwhile, the size effect was also observed in the measured thermal conductivities of nanofilms. Boiko [107] and Nath [108] measured the in-plain thermal conductivities of Al, Ag, and Cu nanofilms using a direct current heating method, and found that the measured thermal conductivity decreased as the film thickness decreased. Kelemen [109], Paddock [110], and Rohde [111] measured the thermal conductivities of Cu, Ni, and Ti nanofilms using optical-thermal and acoustical-thermal methods, and found that the nanostructures (grain size, defects, etc.) inside the nanofilms contributed greatly to the thermal conductivity. Yamane [112], Lee [113] measured the thermal conductivities of Au, Ag Cu, and SiO_2 nanofilms using alternating current heating and 3 ω methods, and found that the measured thermal conductivity was significantly lower than the bulk value when the film thickness was comparable with the electron MFP.

3. Material Properties at Low Temperatures

The non-Fourier heat conduction behaviors normally occur at low temperatures, thus the low temperature material properties should be studied in the first place. Houston [114] introduced a quantized electron scattering mechanism into the Brillouin scattering theory and successfully predicted the temperature dependence of the pure metal resistance as the temperature approaches zero. Gurzhi and Kopeliovich [115] analyzed the motion characteristics of free electrons at Fermi surface at low temperatures, where the electron–phonon interaction was noted to be the main effect. A uniform explanation was given using an electron diffusion equation. Hulm [116] measured the thermal conductivities of Sn, Hg, In, Ta, and their alloys at 1.7–4.3 K, and found that the electron-defect scattering was the dominant factor at low temperatures. Andrews et al [117] measured the thermal conductivity of pure Al at 2–20 K and found a maximum value existed at 14–17 K. Poker and Klabunde [118] measured the electrical conductivities of V, Pt, and Cu at low temperatures, where the experimental data agreed well with the prediction of Bloch-Grüneisen and Wilson theory. Nishi et al. [119] measured the temperature dependence of pure Au and Pb resistances and gave an empirical formula in the middle temperature range.

On the other hand, the study of Wiedemann-Franz (WF) law became important at low temperatures. Kumar et al. [120] summarized a large number of experimental data of electrical and thermal conductivities of different metals, and found that the WF law was valid at low temperatures. Meanwhile, the WF law broke for highly doped cuprate materials, doped MgB_2, quantum dot materials, quasi-one-dimensional organic crystals, etc., [121–129], because these materials had special electronic energy-band structure and Fermi surface shape; the classical Fermi liquid model was not valid anymore. As the material size decreased to nanometers, the WF law could also be violated [130, 131], which was mainly caused by the size effect on the electron scattering mechanisms.

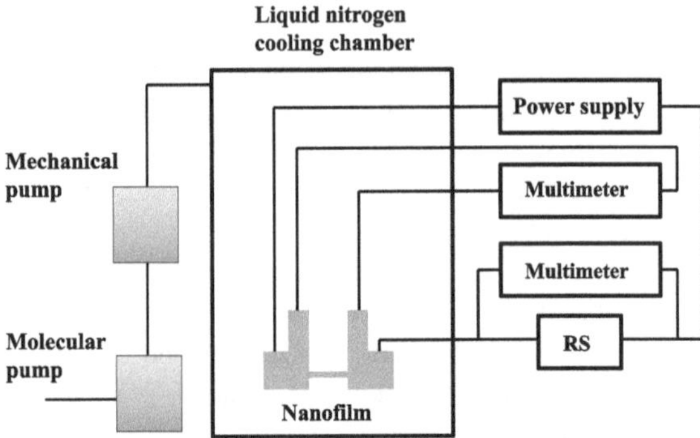

Fig. 1.4 Direct current heating measurement system [132]

4. Direct Current Heating Method

In this thesis, a direct current heating method is used for measurement at low temperatures. Reference [132] gives a scheme diagram of this method. Figure 1.4 shows a typical direct current heating measurement system cooled by liquid nitrogen. The metallic nanofilms are placed in a cryostat kept at a constant temperature and a mechanical pump-molecular pump system is used to maintain a high vacuum environment below 10^{-4} Pa. The film resistance is measured using a four-probe method, and the electrical and thermal conductivities can be extracted simultaneously from the precisely measured resistances at different temperatures; more details can be found in the Chap. 4. In this method, the nanofilm serves as both Joule heater and resistance thermometer, the average temperature of the film is measured.

Our experimental system is similar to the system shown in Fig. 1.4, but a liquid helium cooling system is used instead to provide a minimum temperature of 2.8 K. The main purposes of the experiment are to: (1) obtain the electrical and thermal conductivities of metallic nanofilms at low temperatures; (2) study the effects of different electron scattering mechanisms on the charge and heat transport in nanofilms, test the validation of the WF law for nanofilms; (3) observe the steady non-Fourier heat conduction behaviors in metallic nanofilms under the ultra-high heat flux, low temperature conditions, and prove the validity of the thermomass theory.

1.4 Conclusions

The study of heat conduction under the ultra-short pulsed laser heating, ultra-high heat flux, and very low temperature conditions has become the research front. But some problems remain unsolved at present: (1) transient thermal wave behaviors could only be predicted empirically, no fully developed theory has been proposed to reveal the physical essence. The study of non-Fourier heat conduction in steady states is almost a blank; (2) the experimental research of non-Fourier heat conduction is very rare; (3) the study of electron scattering in metallic nanomaterials at low temperatures is not substantial.

In order to resolve these problems, we focus on the next several aspects to study the non-Fourier heat conduction behaviors:

1. Develop a general heat conduction equation based on the thermomass theory and reveal the physical essence of the non-Fourier heat conduction;
2. Establish a femtosecond laser thermoreflectance system to study the transient energy exchange between the electrons and phonons inside the metallic nanofilms, and measure the electron–phonon coupling factor, interfacial thermal resistance, and other material properties;
3. Establish a direct current heating system at low temperatures, and measure the electrical and thermal conductivities in a wide temperature range. Observe the steady non-Fourier heat conduction phenomenon under the ultra-high heat flux, low temperature conditions, and provide experimental evidences for the thermomass theory.

References

1. J.B. Biot, Memoire sur la propagation de la chaleur. Bibliotheque Brittanique **37**, 310–329 (1804)
2. V. Peshkov, "second sound" in helium II. J Physics-USSR **8**, 381 (1944)
3. L. Landau, Theory of the superfluidity of helium II. Phys. Rev. **60**(4), 356–358 (1941)
4. J.C. Ward, J. Wilks, The velocity of second sound in liquid helium near the absolute zero. Phil. Mag. **42**(326), 314–316 (1951)
5. M. Chester, Second sound in solids. Phys. Rev. **131**(5), 2013–2015 (1963)
6. V. Narayana, R.C. Dynes, Observation of second sound in bismuth. Phys. Rev. Lett. **28**(22), 1461–1465 (1972)
7. S.D. Brorson, J.G. Fujimoto, E.P. Ippen, Femtosecond electronic heat-transport dynamics in thin gold-films. Phys. Rev. Lett. **59**(17), 1962–1965 (1987)
8. C. Cattaneo, Sulla conduzione del calore. Atti Semin. Mat. Fis. Univ. Modena **3**, 83–101 (1948)
9. P. Vernotte, Paradoxes in the continuous theory of the heat equation. C. R. Acad. Sci. **246**, 3154–3155 (1958)
10. P.M. Morse, H. Feshbach, *Methods of Theoretical Physics* (McGraw-Hill, New York, 1953)
11. A. Barletta, E. Zanchini, Hyperbolic heat conduction and local equilibrium: a second law analysis. Int. J. Heat Mass Transf. **40**(5), 1007–1016 (1997)
12. H.S. Chu, W.B. Lor, Hyperbolic heat conduction in thin-film high t_c superconductors with interface thermal resistance. Cryogenics **39**, 739–750 (1999)
13. M.A. Al-Nimr, A.F. Khadrawi, M. Hammad, A generalized thermal boundary condition for the hyperbolic heat conduction model. Heat Mass Transfer **39**, 69–79 (2002)
14. M. Lewandowska, L. Malinowski, An analytical solution of the hyperbolic heat conduction equation for the case of a finite medium symmetrically heated on both sides. Int. Commun. Heat Mass Transfer **33**, 61–69 (2006)
15. D.W. Tang, N. Araki, Analytical solution of non-fourier temperature response in a finite medium under laser-pulse heating. Heat Mass Transfer **31**, 359–363 (1996)
16. B. Pulvirenti, A. Barletta, E. Zanchini, Finite-difference solution of hyperbolic heat conduction with temperature-dependent properties. Numer. Heat Transfer, part A: Applications. **34**(2), 169–183 (1998)
17. Z.M. Tan, W.J. Yang, Heat transfer during asymmetrical collision of thermal waves in a thin film. Int. J. Heat Mass Transfer **40**(17), 3999–4006 (1997)

18. S. Torii, W.J. Yang, Heat transfer mechanisms in thin film with laser heat source. Int. J. Heat Mass Transfer **48**, 537–544 (2005)
19. A. Vedavarz, K. Mitra, S. Kumar, Hyperbolic temperature profiles for laser surface interactions. J. Appl. Phys. **76**(9), 5014–5021 (1994)
20. D.W. Tang, N. Araki, The wave characteristics of thermal conduction in metallic films irradiated by ultra-short laser pulses. J. Phys. D: Appl. Phys. **29**, 2527–2533 (1996)
21. Z.M. Tan, W.J. Yang, Propagation of thermal waves in transient heat conduction in a thin film. J. Franklin Inst. **336B**, 185–197 (1999)
22. K.C. Liu, Numerical simulation for non-linear thermal wave. Appl. Math. Comput. **175**, 1385–1399 (2006)
23. D.Y. Tzou, The generalized lagging response in small-scale and high-rate heating. Int. J. Heat Mass Transf. **38**(17), 3231–3240 (1995)
24. S.J. Su, W.Z. Dai, P.M. Jordan, R.E. Mickens, Comparison of the solutions of a phase lagging heat transport equation and damped wave equation. Int. Heat Mass Transfer **48**, 2233–2241 (2005)
25. S.J. Su, W.Z. Dai, Comparison of the solutions of a phase-lagging heat transport equation and damped wave equation with a heat source. Int. Heat Mass Transfer **49**, 2793–2801 (2006)
26. A. Majumdar, Microscale heat conduction in dielectric thin films. J. Heat Transfer **115**(1), 7–16 (1993)
27. G. Chen, Ballistic diffusive heat conduction equations. Phys. Rev. Lett. **86**(11), 2297–2300 (2001)
28. S.I. Anisimov, B.L. Kapeliovich, T.L. Perelman, Electron emission from surface of metals induced by ultrashort laser pulses. Sov. Phys. JETP **39**, 375–377 (1974)
29. T.Q. Qiu, C.L. Tien, Femtosecond laser heating of multi-layer metals - I. analysis. Int. J. Heat Mass Transfer **37**(17), 2789–2797 (1994)
30. T.Q. Qiu, C.L. Tien, Femtosecond laser heating of multi-layer metals - II. experiments. Int. J. Heat Mass Transfer **37**(17), 2799–2808 (1994)
31. T.Q. Qiu, C.L. Tien, Heat transfer mechanisms during short-pulse laser heating of metals. J. Heat Transfer **115**, 835–841 (1993)
32. R.A. Guyer, J.A. Krumhansl, Solution of the linearized phonon boltzmann equation. Phys. Rev. **148**(2), 766–778 (1966)
33. D.Y. Tzou, *Macro- to Microscale Heat Transfer: The Lagging Behavior* (Taylor & Francis, Washington D C, 1996)
34. G. Khitrova, P.R. Berman, M. Sargent, Theory of pump-probe spectroscopy. J. Opt. Soc. Am. B **5**(1), 160–170 (1988)
35. G.L. Eesley, Observation of nonequilibrium electron heating in copper. Phys. Rev. Lett. **51**(23), 2140–2143 (1983)
36. G.L. Eesley, Generation of nonequilibrium electron and lattice temperatures in copper by picoseconds laser pulses. Phys. Rev. B **33**(4), 2144–2151 (1986)
37. J.G. Fujimoto et al., Femtosecond laser interaction with metallic tungsten and nonequilibrium electron and lattice temperatures. Phys. Rev. Lett. **53**(19), 1837–1840 (1984)
38. H.E. Elsayed-Ali et al., Time-resolved observation of electron-phonon relaxation in copper. Phys. Rev. Lett. **58**(12), 1212–1215 (1987)
39. C. Johnson, *Numerical Solution of Partial Differential Equations by the Finite Element Methods* (Cambridge University Press, New York, 1987)
40. P.B. Allen, Theory of thermal relaxation of electrons in metals. Phys. Rev. Lett. **59**(13), 1460–1463 (1987)
41. H.E. Elsayed-Ali et al., Femtosecond thermoreflectivity and thermotransmissivity ofpolycrystalline and single-crystalline gold films. Phys. Rev. B **43**(5), 4488–4491 (1991)
42. T. Juhasz et al., Time-resolved thermoreflectivity of thin gold films and its dependence on the ambient temperature. Phys. Rev. B **45**(23), 13819–13822 (1992)
43. C.K. Sun et al., Femtosecond-tunable measurement of electron thermalization in gold. Phys. Rev. B **50**(20), 15337–15348 (1994)

44. J. Hohlfeld et al., Time-resolved thermoreflectivity of thin gold films and its dependence on film thickness. Appl. Phys. B: Lasers and Optics **64**(3), 387–390 (1997)
45. N. Taketoshi, T. Baba, O. Akira, Development of a thermal diffusivity measurement system-for metal thin films using a picoseconds thermoreflectance technique. Meas. Sci. Technol. **12**, 2064–2073 (2001)
46. S.D. Brorson, M.K. Kelly, U. Wenschuh, R. Buhleier, J. Kuhl, Femtosecond pump-probe investigation of electron dynamics in idid c_{60} films. Phys. Rev. B **46**(11), 7329–7332 (1992)
47. W.L. McMillan, Transition temperature of strong-coupled superconductors. Phys. Rev. **167**, 331–344 (1968)
48. P.B. Allen, R.C. Dynes, Transition temperature of strong-coupled superconductors reanalyzed. Phys. Rev. B **12**, 905–922 (1975)
49. P.E. Hopkins, J.L. Kassebaum, P.M. Norris, Effects of electron scattering at metal-nonmetal interfaces on electron-phonon equilibration in gold films. J. Appl. Phys. **105**, 023710 (2009)
50. P.E. Hopkins, P.M. Norris, L.M. Phinney, S.A. Policastro, R.G. Kelly, Thermal conductivity in nanoporous gold films during electron-phonon nonequilibrium. J. Nanomater. 418050 (2008)
51. P.E. Hopkins, P.M. Norris, Substrate influence in electron-phonon coupling measurements in thin au films. Appl. Surf. Sci. **253**, 6289–6294 (2007)
52. R.J. Stevens, A.N. Smith, P.M. Norris, Signal analysis and characterization of experimental setup for the transient thermoreflectance technique. Rev. Sci. Instrum **77**, 084901 (2006)
53. R.J. Stevens, A.N. Smith, P.M. Norris, Measurement of thermal boundary conductance of a series of metal-dielectric interfaces by the transient thermoreflectance technique. J. Heat Transfer **127**, 315–322 (2005)
54. P.M. Norris, A.P. Caffrey, R.J. Stevens, J.M. Klopf, J.T. McLeskey Jr, A.N. Smith, Femtosecond pump-probe nondestructive examination of materials. Rev. Sci. Instrum. **74**(1), 400–406 (2003)
55. A.N. Smith, P.M. Norris, Influence of intraband transitions on the electron thermoreflectance response of metals. Appl. Phys. Lett. **78**(9), 1240–1242 (2001)
56. J.T. McLeskey Jr, P.M. Norris, Femtosecond transmission studies of a-si:h, a-sige:h and a-sic:h alloys pumped in the exponential band tails. Sol. Energy Mater. Sol. Cells **69**, 165–173 (2001)
57. W.S. Capinski, H.J. Maris, T. Ruf, M. Cardona, K. Ploog, D.S. Katzer, Thermal conductivity measurements of gaas/alas superlattices using a picosecond optical pump and probe technique. Phys. Rev. B **59**(12), 8105–8113 (1999)
58. T. Nakamiya, T. Ueda, T. Ikegami, F. Mitsugi, K. Ebihara, R. Tsuda, Pulsed laser heating process of multi-walled carbon nanotubes film. Diam. Relat. Mater. **17**, 1458–1461 (2008)
59. S. Wolltersen, U. Emmerichs, H.J. Bakker, Femtosecond mid-ir pump-probe spectroscopy of liquid water: evidence for a two-component structure. Science **278**(24), 658–660 (1997)
60. A. Tokmakoff, B. Sauter, M.D. Fayer, Temperature-dependent vibrational relaxation in polyatomic liquids: picosecond infrared pump-probe experiments. J. Chem. Phys. **100**(12), 9035–9043 (1994)
61. P. Han, D.W. Tang, L.P. Zhou, Numerical analysis of two-dimensional lagging thermal behavior under short-pulse-laser heating on surface. Int. J. Eng. Sci. **44**, 1510–1519 (2006)
62. K. Ramadan, Treatment of the interfacial temperature jump condition with non-fourier heat conduction effects. Int. Commun. Heat Mass Transfer **35**(9), 1177–1182 (2008)
63. Y.M. Lee, T.W. Tsai, Ultra-fast pulse-laser heating on a two-layered semi-infinite material with interfacial contact conductance. Int. Commun. Heat Mass Transfer **34**, 45–51 (2007)
64. D.Y. Tzou, K.S. Chiu, Temperature-dependent thermal lagging in ultrafast laser heating. Int. J. Heat Mass Transfer **44**, 1725–1734 (2001)
65. Y. Yamashita, T. Yokomine, S. Ebara, A. Shimizu, Heat transport analysis for femtosecond laser ablation with molecular dynamics two temperature model method. Fusion Eng. Des. **81**, 1695–1700 (2006)
66. B.H. Christensen, K. Vestentoft, P. Balling, Short-pulse ablation rates and the two temperature model. Appl. Surf. Sci. **253**, 6347–6352 (2007)

67. M.E. Povarnitsyn, T.E. Itina, K.V. Khishchenko, P.R. Levashov, Multi-material two-temperature model for simulation of ultra-short laser ablation. Appl. Surf. Sci. **253**, 6343–6346 (2007)
68. H.J. Wang, W.Z. Dai, L.G. Hewavitharana, A finite difference method for studying thermal deformation in a double-layered thin film with imperfect interfacial contactexposed to ultra-short pulsed lasers. Int. J. Therm. Sci. **47**, 7–24 (2008)
69. H.J. Wang, W.Z. Dai, R. Melnik, A finite difference method for studying thermal deformation in a double-layered thin film exposed to ultrashort pulsed lasers. Int. J. Therm. Sci. **45**, 1179–1196 (2006)
70. H.J. Wang, W.Z. Dai, R. Nassar, R. Melnik, A finite difference method for studying thermal deformation in a thin film exposed to ultrashort-pulsed lasers. Int. J. Heat Mass Transfer **49**, 2712–2723 (2006)
71. A. Saidane, S.H. Pulko, High-power short-pulse laser heating of low dimensional structures: a hyperbolic heat conduction study using tlm. Microelectron. Eng. **51–52**, 469–478 (2000)
72. B.S. Yilbas, A.F.M. Arif, Laser short pulse heating: influence of pulse intensity ontemperature and stress fields. Appl. Surf. Sci. **252**, 8428–8437 (2006)
73. J. Xu, X.W. Wang, Simulation of ballistic and non-fourier thermal transport in ultra-fast laser heating. Phys. B **351**, 213–226 (2004)
74. J.C. Wang, C.L. Guo, Effect of electron heating on femtosecond laser-induced coherent acoustic phonons in noble metals. Phys. Rev. B **75**, 184304 (2007)
75. H.D. Wang, W.G. Ma, X. Zhang, W. Wang, Measurement of thermal wave in metal films using femtosecond laser thermoreflectance system. Acta Physica Sinica **59**(6), 3856–3862 (2010). in Chinese
76. Z.X. Li, X.B. Luo, Z.Y. Guo, Mems technology status and development trend. J. Sens. Technol. **20**(9), 58–60 (2001). in Chinese
77. J.P. Uyemura, *Introduction to Visi Circuits and System Uyemura*, 1st edn. (Wiley, New york, 2001)
78. J.J. Thomson, On the theory of electric conduction through thin metallic films. Proc. Camb. Phil. Soc **11**, 120 (1901)
79. A.C.B. Lovell, Proc. Roy. Soc. (London) **157**, 311 (1936)
80. K. Fuchs, The conductivity of thin metallic films according to the electron theory of metals. Proc. Camb. Phil. Soc. **34**, 100–108 (1938)
81. E.H. Sondheimer, The mean free path of electrons in metals. Advan. Phys. **1**, 1–42 (1952)
82. A.F. Mayadas, M. Shatzkes, J.F. Janak, Electrical resistivity model for polycrystalline films: the case of specular reflection at external surfaces. Appl. Phys. Lett. **14**(11), 345–347 (1969)
83. A.F. Mayadas, M. Shatzkes, Electrical-resistivity model for polycrystalline films: the case of arbitrary reflection at external surfaces. Phys. Rev. B **1**(4), 1382–1389 (1970)
84. C.L. Tien, B.F. Armaly, P.S. Jagannathan, *Thermal conductivity of thin metallic films and wires at cryogenic temperatures. Thermal conductivity*. (Plenum, New york, 1969), pp. 13–19
85. J. Bass, W.P. Pratt, P.A. Schroeder, The temperature dependent electrical resistivities of the alkali metals. Rev. Mod. Phys. **62**(3), 645–744 (1990)
86. Z.S. Chen, X.S. Ge, Y.Q. Gu, *Calorimetry and Determination of Thermal Properties* (University of Science and Technology of China Press, China, 1990). in Chinese
87. Y.Z. Cao, X.G. Qiu, *Experimental Heat Transfer* (National Defense Industry Press, China, 1998). in Chinese
88. X.S. Wang, X.P. Wu, J. Qin et al., Experimental study of the infrared thermal imaging method for measuring the temperature of the flame. Laser and Infrared **3**, 101–104 (2001). in Chinese
89. Z.Q. Yu, C.A. Moore, Y. Hu et al., Measurement of the surface temperature using the laser raman method. Chinese Laser **12**(8), 492–494 (1985). in Chinese
90. S. Paoloni, H.G. Walther, Photothermal radiometry of infrared translucent materials. J. Appl. Phys. **82**(1), 101–106 (1997)
91. G.B. Zhang, J.Y. Shi, C.S. Shi et al., Photoacoustic technology in thermal diffusivity measurements of solid materials. Physics **29**(7), 616–619 (2000). in Chinese

92. E. Doebelin, *Measurement Systems: Application and Design*, 3rd edn. (McGraw-Hill, New York 1985)
93. D.W. Pohl, W. Denk, M. Lanz, Optical stethoscopy: image recording with resolution λ/20. Appl. Phys. Lett. **44**, 651–653 (1984)
94. A. Majumdar, Scanning thermal microscopy. Ann. Rev. Mater. Sci. **29**, 505–585 (1999)
95. C.Y. Bao, W.Y. Feng, X.M. Liu, Laser fluorescence measurement of the gas temperature. J. Tsinghua Univ. (Sci. Technol.) **36**(6), 40–43 (1999). in Chinese
96. B.K. You, *Temperature measurement and instrumentation: thermocouples and thermal resistance* (Science and Technology Literature Publishing House, China, 1990) in Chinese
97. V.P. Duggal, V.P. Nagpal, Size effect in thin single-crystal silver films. Appl. Phys. Lett. **13**(6), 206–207 (1968)
98. V.P. Duggal, V.P. Nagpal, Geometrical size effect in resistivity and hall coefficient in single-crystal silver films. J. Appl. Phys. **42**(11), 4500–4502 (1971)
99. L.R. Kirkland, R.L. Chaplin, Electrical size effect of aluminum single crystals. J. Appl. Phys. **42**(8), 3054–3057 (1971)
100. L.A. Moraga, J. Caballero, G. Kremer, Electrical resistivity of very thin single-crystal titanium films as a function of temperature. Thin Solid Films **117**, 1–8 (1984)
101. G. Kästle, H.G. Boyen, A. Schröder et al., Size effect of the resistivity of thin epitaxial gold films. Phys. Rev. B **70**(16), 165414 (2004)
102. G. Ramaswamy, A.K. Raychauhuri, J. Goswami et al., Scanning tunneling microscope study of the morphology of chemical vapor deposited copper films and its correlation with resistivity. J. Appl. Phys. **82**(8), 3797–3807 (1997)
103. M. Fenn, G. Akuetey, P.E. Donovan, Electrical resistivity of cu and nb thin films. J. Phys.: Condens. Matter **10**, 1707–1720 (1998)
104. C. Durkan, M.E. Welland, Size effects in the electrical resistivity of polycrystalline nanowires. Phys. Rev. B **61**(20), 14215–14218 (2000)
105. W. Wu, S.H. Brongersma, M. Van Hove et al., Influence of surface and grain-boundary scattering on the resistivity of copper in reduced dimensions. Appl. Phys. Lett. **84**(15), 2838–2840 (2004)
106. H. Maroma, M. Eizenberg, The effect of surface roughness on the resistivity increase in nanometric dimensions. J. Appl. Phys. **99**(12), 123705 (2006)
107. B.T. Boiko, A.T. Pugachev, V.M. Bratsychin, Method for the determination of the thermophysical properties of evaporated thin films. Thin Solid Films **17**, 157–161 (1973)
108. P. Nath, K.L. Chopra, Thermal conductivity of copper films. Thin Solid Films **20**, 53–62 (1974)
109. F. Kelemen, Pulse method for the measurement of the thermal conductivity of thin films. Thin Solid Films **36**, 199–203 (1976)
110. C.A. Paddock, G.L. Eesley, Transient thermoreflectance from thin metal films. J. Appl. Phys. **60**(1), 285–290 (1986)
111. M. Rohde, Photoacoustic characterization of thermal transport properties in thin films and microstructures. Thin Solid Films. **238**, 199–206, (1994)
112. T. Yamane, Y. Mori, S. Katayama et al., Measurement of thermal diffusivities of thin metallic films using the ac calorimetric method. J. Appl. Phys. **82**(3), 1153–1156 (1997)
113. S.M. Lee, D.G. Cahill, Heat transport in thin dielectric films. J. Appl. Phys. **81**(6), 2590–2595 (1997)
114. W.V. Houston, The temperature dependence of electrical conductivity. Phys. Rev. **34**, 279–283 (1929)
115. R.N. Gurzhi, A.I. Kopeliovich, Low-temperature electrical conductivity of pure metals. Sov. Phys. Usp. **24**(1), 17–41 (1981)
116. J.K. Hulm, The thermal conductivity of tin, mercury, indium and tantalum at liquid helium temperatures. Proc. R. Soc. Lond. A **204**, 98–123 (1950)
117. F.A. Andrews, R.T. Webber, D.A. Spohr, Thermal conductivities of pure metals at low temperatures. I. Alum. Phys. Rev. **84**(5), 994–996 (1951)

118. D.B. Poker, C.E. Klabunde, Temperature dependence of electrical resistivity of vanadium, platinum and copper. Phys. Rev. B **26**(12), 7012–7014 (1982)
119. Y. Nishi, A. Igarashi, K. Mikagi, Temperature dependence of electrical resistivity for gold and lead. J. Mater. Sci. Lett. **6**, 87–88 (1987)
120. G.S. Kumar, G. Prasad, R.O. Pohl, Review, experimental determinations of the lorenz number. J. Mater. Sci. **28**, 4261–4272 (1993)
121. A. Houghton, S. Lee, J.B. Marston, Violation of the wiedemann-franz law in a large-n solution of the $t-j$ model. Phys. Rev. B **65**, 220503 (2002)
122. A.V. Sologubenko, N.D. Zhigadlo, J. Karpinski, H.R. Ott, Thermal conductivity of al-doped mgb_2: impurity scattering and the validity of the wiedemann-franz law. Phys. Rev. B **74**, 184523 (2006)
123. M.G. Vavilov, A.D. Stone, Failure of the wiedemann-franz law in mesoscopic conductors. Phys. Rev. B **72**, 205107 (2005)
124. G.Z. Liu, G. Cheng, Chiral symmetry breaking and violation of the wiedemann franz law in underdoped cuprates. Phys. Rev. B **66**, 100505 (2002)
125. M.F. Smith, R.H. McKenzie, Apparent violation of the wiedemann-franz law near a magnetic field tuned metal-antiferromagnetic quantum critical point. Phys. Rev. Lett. **101**, 266403 (2008)
126. K. Vafayi, M. Calandra, O. Gunnarsson, Electronic thermal conductivity at high temperatures: violation of the wiedemann-franz law in narrow-band metals. Phys. Rev. B **74**, 235116 (2006)
127. A. Casian, Violation of the wiedemann-franz law in quasi-one-dimensional organic crystals. Phys. Rev. B **81**, 155415 (2010)
128. K.S. Kim, C. Pépin, Violation of the wiedemann-franz law at the kondo breakdown quantum critical point. Phys. Rev. Lett. **102**, 156404 (2009)
129. A. Garg, D. Rasch, E. Shimshoni, A. Rosch, Large violation of the wiedemann franz law in luttinger liquids. Phys. Rev. Lett. **103**, 096402 (2009)
130. N. Stojanovic, D.H.S. Maithripala, J.M. Berg, M. Holtz, Thermal conductivity in metallic nanostructures at high temperature: electrons, phonons, and the wiedemann franz law. Phys. Rev. B **82**, 075418 (2010)
131. Q.G. Zhang, B.Y. Cao, X. Zhang, M. Fujii, K. Takahashi, Influence of grain boundary scattering on the electrical and thermal conductivities of polycrystalline gold nanofilms. Phys. Rev. B **74**, 134109 (2006)
132. Q.G. Zhang, B.Y. Cao, X. Zhang, M. Fujii, K. Takahashi, Size effects on the thermal conductivity of polycrystalline platinum nanofilms. J. Phys.: Condens Matter. **18**, 7937–7950 (2006)

Chapter 2
Thermomass Theory for Non-Fourier Heat Conduction

Abstract With the rapid development of femtosecond laser and micro/nano processing techniques, researchers face great challenges in thermal management and analysis under the extreme conditions. As the theoretical basis of heat transfer, Fourier's law may break down at femtosecond temporal scales and nanometer spatial scales. In 1822, Fourier stated in his book "Analytical theory of heat" that the mechanical principles could not be applied to study the thermal phenomenon, which used concepts that differed from other fields of study [1]. But the heat transport in metals can be analogous to the charge transport according to the Wiedemann–Franz (WF) law [2], it shows internal connection between thermal science and other branches of physics. Guo has developed a novel thermomass theory to analyze the heat conduction using Newtonian mechanics [3], creating a new path for thermal analysis under the extreme conditions.

2.1 Definition of Thermomass and the State Equation of Thermon Gas

2.1.1 Definition of Thermomass and Themon Gas

Thermomass (TM) is defined based on Einstein's mass–energy relation [3, 4], it is actually the relativistic mass of thermal energy. The mass–energy relation is given as:

$$E = Mc^2 = \frac{M_0 c^2}{\sqrt{1 - v^2/c^2}} \tag{2.1}$$

where E, M, v, and c are the thermal energy, rest mass, velocity, and speed of light, respectively. M is the relativistic mass. When the velocity is far less than the speed of light, the Eq. (2.1) can be simplified as:

H.-D. Wang, *Theoretical and Experimental Studies on Non-Fourier Heat Conduction Based on Thermomass Theory*, Springer Theses, DOI: 10.1007/978-3-642-53977-0_2,

$$E = Mc^2 = \left(M_0 + \frac{1}{2}\frac{M_0 v^2}{c^2} + \cdots\right)c^2 \approx (M_0 + M_k)\,c^2 \tag{2.2}$$

where M_k is the additive mass induced by kinetic energy. The TM M_h is the relativistic mass of internal energy U:

$$M_h = \frac{U}{c^2} \tag{2.3}$$

The total mass M_t can be written as the sum of the rest mass, relativistic mass of kinetic energy and TM:

$$M_t = M_0 + M_k + M_h \tag{2.4}$$

If $v = 0$, the total mass equals the sum of rest mass and TM. The density of TM is given as:

$$\rho_h = \frac{\rho C_v T}{c^2} \tag{2.5}$$

where ρ, C_v and T are the density, specific heat per volume, and temperature, respectively. The drift velocity of TM is given as:

$$u_h = \frac{q}{\rho C_v T} \tag{2.6}$$

where q is the heat flux.

It has to be emphasized that the concept of "thermomass" is different from the concept "caloric" in eighteenth century. Caloric was known as an imaginary, massless fluid that could be neither created nor destroyed. Lavoisier called caloric as the substance of heat, which was even regarded as a new type of "chemical element" [5]. Many thermal phenomena could be explained by the caloric theory, e.g., hot bodies had more caloric while cold ones had less, the flow of caloric from hotter to colder bodies formed heat flux. In the early nineteenth century, Carnot developed his famous principle of the Carnot cycle solely from the caloric viewpoint, and this principle still forms the basis of heat engine theory today [6]. But the caloric theory could not be used to explain the heat generation by friction, since there was nothing but motion communicated to the body that was heated. This problem was successfully resolved by a kinetic theory that explained heat as random movement of particles (atoms, molecules) of the substance. The kinetic theory then became the foundation of modern thermodynamics [7].

The TM is defined as the relativistic mass of thermal energy, unlike the imaginary fluid of caloric. The TM theory reveals a dual nature of heat, i.e., heat behaves like energy during its transformation into other forms of energy; heat behaves like mass during its transfer from hotter to colder bodies [3]. This is much like the wave particle duality of light. Actually, photons exhibit wave properties during their propagation, while they exhibit particle properties during their interaction with other particles, such as electrons and phonons. Einstein's mass–energy relation demonstrates that mass and energy are two names for the same thing. Neither one can exist without

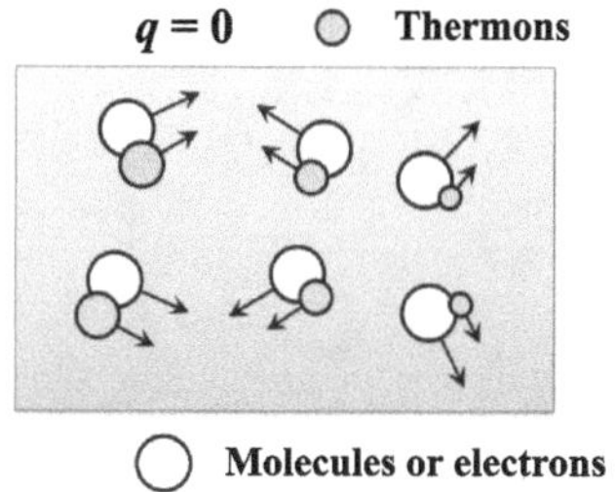

Fig. 2.1 Thermon gas at a uniform temperature

the other, they are two properties connected by a constant. The TM theory is the application example of mass–energy relation in thermal science.

In fact, the mass nature of heat has already been noticed by other researchers in the last century [8]. In 1931, Onsager [9] pointed out in his study of irreversible thermodynamics that Fourier's law was just an approximation of the heat conduction process, ignoring the time needed for the acceleration of the heat flow. Thus Fourier's law went against the principle of microreversibility in thermodynamics. The inertia or lagging effect between "force" and "flow" commonly exist in mechanics and electromagnetic. In thermal science, the TM is usually very small and the corresponding inertia effect is negligible, the heat conduction behaves like a diffusion process. In modern physics, the TM inertia may not be negligible under the ultra-fast laser heating and ultra-high heat flux conditions, leading to non-Fourier heat conduction. A Newtonian mechanics is used in the TM theory to establish a general heat conduction law to replace Fourier's law under these extreme conditions. From a microscopic point of view, the TM of an individual particle (atoms, molecules, etc.) is defined as "Thermon," all the thermons in a system form thermon gas. The heat conduction of solid is actually the direct flow of thermon gas under its pressure gradient [10].

Figure 2.1 shows a system filled with air molecules or electrons (white circles), the thermons attached to the particles are shown as the yellow circles. At a uniform temperature, all the particles are in the state of thermal equilibrium. The particles and thermons move randomly and the average velocity is zero, the heat flux is also zero.

Figure 2.2 shows a system that is hot on the left side and cold on the right side. Because the system is closed, the average velocity of rest mass particles is still zero. But the thermon gas moves from the hot side to the cold side driven by its pressure gradient, which is proportional to the temperature gradient. A heat flux is established accordingly. The density of thermon gas is $\rho_h = \rho C_v T/c^2$, its drift velocity is $u_h = q/(\rho C_v T)$. In order to analyze the motion of thermon gas theoretically, its state equations of different materials should be given first.

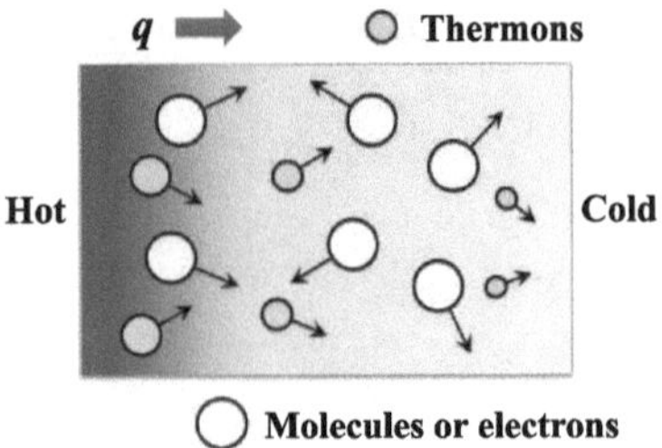

Fig. 2.2 Thermon gas flow driven by a temperature gradient

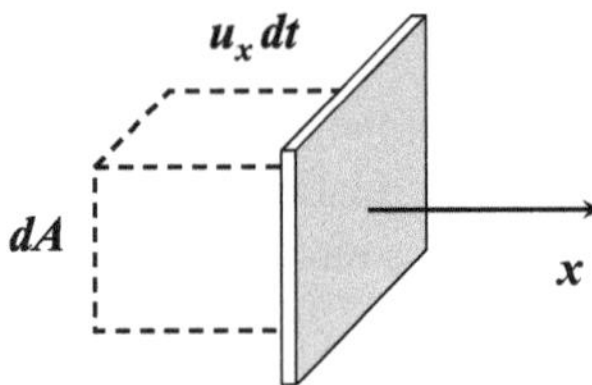

Fig. 2.3 Cube element for calculating the impulse

2.1.2 State Equation of Thermon Gas in Ideal Gas

There are several assumptions for the thermon gas in ideal gas: (1) the thermons are attached to the air molecules and satisfy the classical Maxwell–Boltzmann distribution function; (2) Newtonian mechanics is applicable for the thermon gas. For a system filled with randomly moving particles, the impulse is given as:

$$i = mu_x \tag{2.7}$$

where m, i and u_x are the particle mass, impulse, and velocity in x direction, respectively. A cube at area element dA perpendicular to x direction has a volume of $u_x dtdA$ as shown in Fig. 2.3.

The total impulse of the particles in the cube element is:

$$dI = Ni = inu_x dtdA = nmu_x^2 dtdA \tag{2.8}$$

where $N = nu_x dtdA$ is the number of particles, n is the number density. According to the relation between impulse and pressure $dI = PdAdt$, the pressure can be obtained as:

$$P = nmu_x^2 \tag{2.9}$$

Given a consideration to the spatial symmetry in x, y, and z directions, the square of velocity is given as:

$$u_x^2 = u_y^2 = u_z^2 = \frac{1}{3}\bar{u}^2 \tag{2.10}$$

The pressure is then given as $P = \frac{1}{3}nm\bar{u}^2$ and the pressure of thermon gas is:

$$P_h = \frac{1}{3}nm_h\bar{u}^2 = \frac{1}{3}n\bar{u}^2\left(\frac{1}{2}\frac{m\bar{u}^2}{c^2}\right) = \frac{1}{6}\frac{nm\bar{u}^4}{c^2} \tag{2.11}$$

The classical Maxwell–Boltzmann distribution function is:

$$f_M(u) = 4\pi u^2\left(\frac{m}{2\pi k_B T}\right)^{3/2}\exp\left(-\frac{mu^2}{2k_B T}\right) \tag{2.12}$$

where k_B is the Boltzmann constant.

$$\bar{u}^4 = \int_0^\infty u^4 f_M(u)du = 15\left(\frac{k_B T}{m}\right)^2 \tag{2.13}$$

Substituting the Eq. (2.13) into the Eq. (2.11), one can get:

$$P_h = \frac{5}{3}\frac{(r-1)}{c^2}\rho C_v^2 T^2 = \frac{5}{3}\frac{\rho C_v R T^2}{c^2} \tag{2.14}$$

where R is the ideal gas constant, r is the ratio of specific heat capacities at constant pressure and constant volume. The Eq. (2.14) shows that the thermon gas pressure is proportional to the square of temperature and the pressure gradient is the driven force of thermon gas.

2.1.3 State Equation of Thermon Gas in Dielectrics

In dielectrics, the phonons are the quantized quasiparticles of the vibration modes of elastic structures, they are the main energy carriers. The direct flow of phonon gas forms heat flux, the phonons can be seen as the thermons for dielectrics. The unit of driven force of thermon gas is Newton, this is quite different from the other generalized forces in irreversible thermodynamics [1, 11]. The Debye state equation for dielectrics is [4, 12]:

$$P = -\frac{\partial E}{\partial V} + \frac{\gamma \partial E_D}{V} \tag{2.15}$$

where E_D is the energy of lattice thermal vibration, γ is the Grüneisen parameter, describes the effect that changing temperature has on the size or dynamics of the lattice. When the temperature is higher than the Debye temperature, the lattice thermal vibration equation is:

$$P_0 V = \gamma E_{D0} = 3\gamma M_0 R T = \gamma M_0 C_v T \tag{2.16}$$

where E_D is the energy of the rest mass of lattices, the TM can be defined as $M_h = E_D/c^2$. Then the total energy of lattice thermal vibration is:

$$E_D = E_{D0} + E_h = (M_0 + M_h)\, C_v T \tag{2.17}$$

where E_h is the energy of the TM. So the state equation of thermon gas in dielectrics can be given as:

$$P_h V = \gamma M_h C_v T \tag{2.18}$$

$$P_h = \gamma \rho_h C_v T = \gamma \frac{\rho C_v T}{c^2} C_v T = \frac{\gamma \rho}{c^2} (C_v T)^2 \tag{2.19}$$

The pressure of thermon gas P_h is proportional to the square of temperature, the same as in ideal gases. For Si at room temperature, P_h is about 5×10^{-3} Pa.

2.1.4 State Equation of Thermon Gas in Metals

In metals, electrons are the main carriers for charge and heat, some basic assumptions of the classical Drude–Lorentz electron gas model are [13]: (1) the free electron gas in metals behaves like ideal gas; (2) the potential energy between electrons can be neglected and the Maxwell–Boltzmann distribution function is applicable; (3) a thermal equilibrium state is established between the electrons and lattices by collisions, an electron relaxation time τ_e represents the collision time interval.

The Drude–Lorentz model successfully predicts the Wiedemann–Franz (WF) law, but also results in some disagreement with the experimental observation. For example, the specific heat of electron gas can be calculated as: $C_{ve} = \left(\frac{\partial u}{\partial T}\right)_v = 1.5 n k_B \approx 10^6\,\text{Jm}^{-3}\text{K}^{-1}$ according to the Drude–Lorentz model, this value is two orders of magnitudes larger than the experimental value. The interactions between electrons and ions should be taken into account to get a satisfied result. Later, Sommerfeld introduced a three-dimensional potential box with an infinite barrier at the surfaces into the classical free electron gas model, the potential energy was taken as a constant within the box. According to the Pauli exclusion principle, a Fermi–Dirac statistics was applied in the Sommerfeld model matching well with the experimental data. Here, the Sommerfeld model is used to establish the state equation of thermon gas.

The thermons in metals are attached to the electrons and the Fermi–Dirac distribution function should be used. The pressure of thermon gas is given as:

$$P_h = \frac{1}{3} n m_h u_h^2 \tag{2.20}$$

where the TM $m_h = \frac{\varepsilon}{c^2}$, ε is the internal energy including contributions of both electrons and lattices. $u_h = \sqrt{\frac{2\varepsilon}{m}}$ is the velocity of randomly moving particles, thus

$P_h = \frac{2}{3}\frac{n\varepsilon^2}{c^2 m}$. At a thermal equilibrium state, the electron number between ε and $\varepsilon + d\varepsilon$ is:

$$dn = \sigma(\varepsilon, T)d\varepsilon = f(\varepsilon, T)Z(\varepsilon)d\varepsilon \tag{2.21}$$

The Fermi–Dirac distribution function is:

$$f(\varepsilon, T) = \frac{1}{\exp\left(\frac{\varepsilon - \varepsilon_F}{k_B T}\right) + 1} \tag{2.22}$$

The Sommerfeld electron state density function is:

$$Z(\varepsilon) = \frac{1}{2\pi^2}\left(\frac{2m}{\hbar^2}\right)^{\frac{3}{2}} \varepsilon^{\frac{1}{2}} \tag{2.23}$$

where ε_F, k_B and $\hbar$ are the Fermi energy level, Boltzmann constant, and Planck constant, respectively. Further, the pressure of thermon gas is given as:

$$P_h = \frac{2}{3mc^2}\int_0^{\infty} \varepsilon^2 f(\varepsilon, T)Z(\varepsilon)d\varepsilon \tag{2.24}$$

The internal energy of metals is mainly contained in lattices. Due to the strong electron–lattice coupling, the most internal energy is carried by electrons rather than lattice vibration. m in the Eq. (2.24) is responsible for the internal energy, which is mainly decided by the lattices. At the high temperature limit, the energy of lattices is $E = 3k_B T$, thus:

$$nmC_{\max}T = 3nk_B T \tag{2.25}$$

The maximum mass $m_{\max}$ is:

$$m_{\max} = \frac{3k_B}{C_{\max}} \tag{2.26}$$

where $C_{\max}$ is the maximum value of specific heat of metals. The temperature-dependent specific heat of gold is given in the Fig. 2.4.

Combining the lattice dynamics theory [14] and existed experimental data [15, 16], an empirical formula of the specific heat of gold is given as:

$$C = \begin{cases} [T < 0.1\theta_D]: & \gamma T + \beta T^3 \\ [0.1\theta_D < T < 280\,\text{K}]: & 9nk_B\left(\frac{T}{\theta}\right)^3 \int_0^{\theta/T} \frac{x^4 e^x}{(e^x - 1)^2}\,dx \\ [280\,\text{K} < T]: & C_{\max} \end{cases} \tag{2.27}$$

where γ and β are the empirical parameters, θ is the Debye temperature. When the temperature is higher than 300 K, the specific heat C reaches $C_{\max}$. At low temperatures, the mass m is given as:

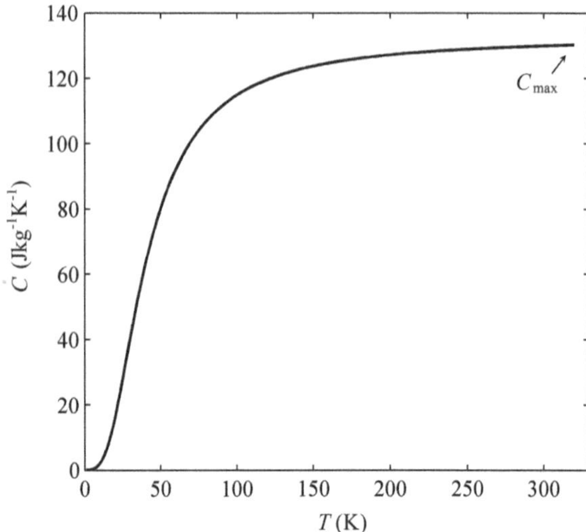

Fig. 2.4 Cube element for calculating the impulse

$$m = \frac{Cm_{\max}}{C_{\max}} = \frac{\rho C_v}{3nk_B} \frac{3\rho k_B}{3nk_B} = \frac{\rho^2 C_v}{3n^2 k_B} \tag{2.28}$$

Substitute the Eq. (2.28) into the Eq. (2.24) and integrate, one can get:

$$\begin{aligned} P_h &\cong \frac{2}{3mc^2}\left[\frac{1}{7\pi^2}\left(\frac{2m_e}{\hbar^2}\right)^{\frac{3}{2}} \varepsilon_F^{7/2} + \frac{5}{24}(k_B T)^2 \left(\frac{2m_e}{\hbar^2}\right)^{\frac{3}{2}} \varepsilon_F^{3/2}\right] \\ &= C_{\varepsilon_F} + \frac{5}{36}\frac{k_B^2}{c^2 m}\left(\frac{2m_e}{\hbar^2}\right)^{\frac{3}{2}} \varepsilon_F^{3/2} T^2 \end{aligned} \tag{2.29}$$

where the first term represents the contribution of electron energy to the TM pressure when $T = 0\,\text{K}$; the second term represents the thermal contribution when $T > 0\,\text{K}$. Because the temperature gradient makes the driven force for heat flux, the temperature-dependent part is written as P_{hT}:

$$P_{hT} = \frac{5}{36}\frac{k_B^2}{c^2 m}\left(\frac{2m_e}{\hbar^2}\right)^{\frac{3}{2}} \varepsilon_F^{3/2} T^2 \tag{2.30}$$

where m_e is the electron mass. The Fermi energy level is $\varepsilon_F = \frac{\hbar^2}{2m_e}\left(3\pi^2 n\right)^{\frac{2}{3}}$ and the Eq. (2.30) can be simplified as:

$$P_{hT} = \frac{5}{12}\frac{\pi^2 n k_B^2}{c^2 m} T^2 \tag{2.31}$$

Substituting the Eq. (2.28) into the Eq. (2.31), one can get:

$$P_{hT} = \frac{5}{12}\frac{\pi^2 nk_B^2}{c^2 m}T^2 = \frac{5}{4}\frac{\pi^2 n^3 k_B^3 T^2}{\rho^2 C_v c^2} \tag{2.32}$$

Compare the Eq. (2.32) with the Eqs. (2.14) and (2.19), one can find that the pressure of thermon gas is proportional to the square of temperature with different coefficients due to the different distribution functions.

2.1.5 Unified State Equation of Thermon Gas

According to the quantum mechanics, the elementary particles include fermions and bosons [17]. Fermions are the particles that have odd half-integer spins; bosons are the ones that have integer spins. Electrons, positrons, neutrons, and protons are all fermions; phonons and photons are bosons. The different elementary classes of phonons and electrons explain the different expressions of thermon gas in dielectrics and metals. But from a macroscopic point of view, these two systems have something in common: (1) thermons have no rest mass in both systems; (2) at a uniform temperature, the thermons are attached to the rest mass particles and its pressure is decided by the different distribution functions of particles; (3) the rest mass of particles keeps constant during collisions, but the TM changes; (4) the motion of thermon gas forms heat flux in both systems, the energy conservation is presented as the mass conservation of thermon gas. These common facts imply that the thermon gas behaves similarly in dielectrics and metals, the difference disappears at high temperature limit and a unified state equation exists.

At high temperatures, the mass m in the Eq. (2.31) is $3k_B/C_v$ and the pressure of thermon gas is:

$$P_{hT} = \frac{5}{36}\frac{\pi^2 nk_B C_v}{c^2}T^2 \tag{2.33}$$

The volume of thermon gas is:

$$V_h = \frac{m_{\max}}{\rho_h} = \frac{3k_B c^2}{\rho C_v^2 T} \tag{2.34}$$

The Grüneisen parameter is:

$$\gamma_h = \frac{P_{hT}V_h}{E} = \frac{P_{hT}V_h}{3k_B T} = \frac{5}{36}\frac{\pi^2 nk_B}{\rho C_v} \tag{2.35}$$

Substituting the Eq. (2.35) into the Eq. (2.33), one can get:

$$P_{hT} = \frac{\gamma_h \rho C_v^2 T^2}{c^2} \tag{2.36}$$

It is noted that the Eq. (2.36) is the same as the Eq. (2.19), it is the unified state equation of thermon gas. At low temperatures, the difference of statistics of different micro-particle systems play an important role and causes a deviation between state equations.

Another question is how to decide the Grüneisen parameter? For metals, the electrons are fermions, the particle number N, internal energy E and pressure P are given as [17]:

$$N = \frac{\sqrt{2}V}{\pi^2}\left(\frac{m}{\hbar^2}\right)^{3/2}\int_0^\infty \frac{\varepsilon^{0.5}d\varepsilon}{1+\exp\left[(\varepsilon-\mu)/(k_B T)\right]} \tag{2.37}$$

$$E = \frac{\sqrt{2}V}{\pi^2}\left(\frac{m}{\hbar^2}\right)^{3/2}\int_0^\infty \frac{\varepsilon^{1.5}d\varepsilon}{1+\exp\left[(\varepsilon-\mu)/(k_B T)\right]} \tag{2.38}$$

$$PV = \frac{2\sqrt{2}V}{3\pi^2}\left(\frac{m}{\hbar^2}\right)^{3/2}\int_0^\infty \frac{\varepsilon^{1.5}d\varepsilon}{1+\exp\left[(\varepsilon-\mu)/(k_B T)\right]} \tag{2.39}$$

For a fermion system, the relation between pressure and internal energy is $PV = 2/3E$, thus the Grüneisen parameter can be calculated as:

$$\gamma = \frac{PV}{E} = \frac{2}{3} \tag{2.40}$$

But the Eq. (2.40) is only valid under the high temperature, high pressure conditions, normally the Grüneisen parameter will differ from 2/3. Grüneisen parameter expresses the effect that changing temperature has on the pressure. The electrons and phonons are strongly coupled in metals, so the contribution of lattices to the Grüneisen parameter of electron gas should be also taken into consideration.

2.2 Non-Fourier Heat Conduction Equation in Unsteady States

2.2.1 Governing Equation of Motion of Thermon Gas

Newtonian mechanics can be applied to analyze the motion of thermon gas. One-dimensional conservation equations of mass and momentum for thermon gas are given as:

$$\frac{\partial \rho_h}{\partial t} + \frac{\partial}{\partial x}(\rho_h u_h) = \frac{S}{c^2} \tag{2.41}$$

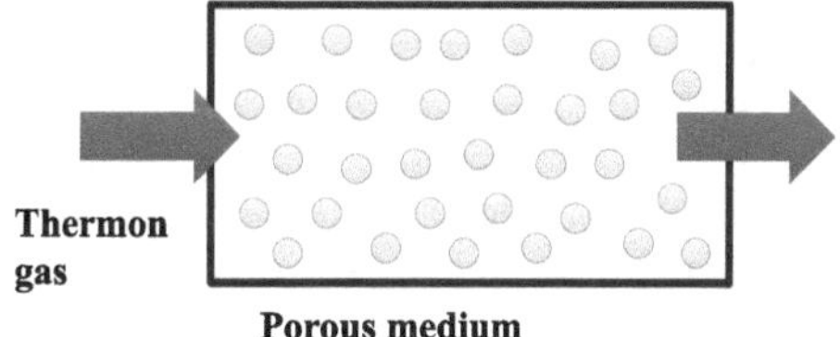

Fig. 2.5 Schematic diagram of the flow of thermon gas in solid

$$\frac{\partial}{\partial t}(\rho_h u_h) + \frac{\partial}{\partial x}(u_h \cdot \rho_h u_h) + \frac{\partial P_h}{\partial x} + f_h = 0 \tag{2.42}$$

where S is the internal heat source, f_h is the resistance. The Eq. (2.41) is actually the energy conservation equation. Substituting the Eq. (2.41) into the Eq. (2.42), one can get:

$$\rho_h \left(\frac{\partial u_h}{\partial t} + u_h \frac{\partial u_h}{\partial x} \right) + \frac{u_h S}{c^2} + \frac{\partial P_h}{\partial x} + f_h = 0 \tag{2.43}$$

The flow of thermon gas in solid can be seen as the flow of compressible fluid in porous medium, where the molecules, atoms, or lattices in solid form the framework of the porous medium as shown in the Fig. 2.5.

Darcy's law can be applied to analyze the motion of thermon gas in solid [18]:

$$f_h = \beta u_h \tag{2.44}$$

where β is the coefficient of TM resistance that should be decided by the experiment. The resistance of thermon gas is proportional to its drift velocity.

2.2.2 General Heat Conduction Equation

A general heat conduction law can be established based on the TM theory from the first principles. This general law reveals the inertia effect of TM in heat conduction, just like the fluid inertia in hydrodynamics, vibration inertia in mechanical oscillators, electromagnetic inertia in electrodynamics, etc. The inertia effect roots in the fundamental physical principles that a physical object resists any change in its state of motion. The inertia occurs commonly between any kinds of generalized force and flow. The TM theory reveals that heat has a nature of mass, which is too small to be observed under the common conditions, but under the extreme conditions of ultra-fast heating or ultra-high heat flux, the inertia of TM will represent itself and cause detectable influences to the heat conduction process.

Combining the state equation of thermon gas, the momentum conservation equation of thermon gas 2.43 can be transformed into:

$$\tau_h\left(\frac{\partial q}{\partial t}-\frac{q}{T}\frac{\partial T}{\partial t}+\frac{q}{\rho C_v T}\frac{\partial q}{\partial x}-\frac{q^2}{\rho C_v T^2}\frac{\partial T}{\partial x}+\frac{q}{\rho C_v T}S\right)+\kappa\frac{\partial T}{\partial x}+q=0 \quad (2.45)$$

where κ, S and τ_h are the thermal conductivity, internal heat source, and characteristic time of thermon gas, respectively. For dielectrics, τ_h can be expressed as:

$$\tau_h=\frac{\kappa}{2\gamma_h \rho C_v^2 T} \quad (2.46)$$

Kinetic theory gives a relaxation time for heat conduction process as [19]:

$$\tau=\frac{3\kappa}{\rho C_v v_s^2}=\frac{\kappa}{\rho C_v v_{tw}^2} \quad (2.47)$$

where v_s is the sound speed and v_{tw} is the propagation speed of thermal wave, $v_{tw}=(v_s/3)^{1/2}$. If one substitutes the classical expression of sound speed $v_s=[r(r-1)C_v T]^{1/2}$ into the Eq. (2.47), it gives:

$$\tau=\frac{3\kappa}{r(r-1)\rho C_v^2 T} \quad (2.48)$$

where r is the ratio of specific heats. Compare the Eqs. (2.48) and (2.46), one can find that the characteristic times of thermon gas and ideal gas have similar expressions, only differing in the constant coefficient. But the physical meanings of τ_h and τ are quite different, τ_h represents the time delay between the flow of thermon gas and its driven force, while τ represents the time needed to return to the equilibrium state from the nonequilibrium state.

Substituting the Eq. (2.41) into the Eq. (2.45), one can get:

$$\tau_h\left(\frac{\partial q}{\partial t}+2u_h\frac{\partial q}{\partial x}-u_h^2\rho C_v\frac{\partial T}{\partial x}\right)+\kappa\frac{\partial T}{\partial x}+q=0 \quad (2.49)$$

The energy conservation equation holds as:

$$\frac{\partial q}{\partial x}=-\rho C_v\frac{\partial T}{\partial t}+S \quad (2.50)$$

To replace q in the Eq. (2.49) with T, the governing equation for temperature can be obtained as:

$$\rho C_v\tau_h\frac{\partial^2 T}{\partial t^2}+\rho C_v\frac{\partial T}{\partial t}=\left(\kappa-\tau_h\frac{q^2}{\rho C_v T^2}\right)\frac{\partial^2 T}{\partial x^2}-\frac{2\tau_h q}{T}\frac{\partial^2 T}{\partial t\partial x}+\tau_h\frac{\partial S}{\partial t}+\frac{2\tau_h q}{\rho C_v T}\frac{\partial S}{\partial x}+S \quad (2.51)$$

The Eq. (2.51) is the general heat conduction equation (law). Obviously, it is a damped wave equation for temperature, the first term on the left side of the equal sign is caused by the temporal inertia of TM, the first two terms on the right side of the equal sign are caused by the spatial inertia of TM. If $\tau_h = 0$, the general equation 2.51 will reduce to Fourier's heat diffusion equation. In other words, the non-Fourier heat conduction behavior is the consequence of TM inertia. In unsteady states, the general heat conduction equation can be used to predict the propagation of thermal wave, while Fourier's equation results in a paradox of infinite propagation speed of thermal disturbance. Furthermore, the general heat conduction law predicts that the non-Fourier heat conduction phenomenon exists in steady states for the first time.

2.2.3 Two-Step Thermomass Model for Metals

When the metals are heated by ultra-short pulsed lasers, the heat conduction process can be divided into two steps [20]: (1) the electrons absorb the laser energy and then transfer the energy to the lattices by electron–phonon coupling; (2) after reaching the equilibrium temperature of electron–phonon system, the energy of lattices is dissipated by heat diffusion. The first step happens within 10 ps and the thermal equilibrium between electrons and lattices can be established around tens of picoseconds. Take this two-step model into consideration, a two-step TM model (TSTM) has been developed. Several assumptions should be satisfied for TSTM: (1) the state equations of thermon gas in dielectrics and metals are applicable for lattice and electron systems, respectively; (2) ignore the scattering effect at defects and grain boundaries of the metals; (3) the electron–phonon collisions are presented by an electron–phonon coupling factor G.

For electron system, the mass and momentum conservation equations are listed as:

$$\frac{\partial \rho_h}{\partial t} + \rho_h \frac{\partial u_h}{\partial x} + u_h \frac{\partial \rho_h}{\partial x} = \frac{S - G(T_e - T_l)}{c^2} \tag{2.52}$$

$$\rho_h \left(\frac{\partial u_h}{\partial t} + u_h \frac{\partial u_h}{\partial x} \right) + \frac{u_h}{c^2} [S - G(T_e - T_l)] + \frac{\partial P_h}{\partial x} + f_h = 0 \tag{2.53}$$

where T_e, T_l are the electron temperature and lattice temperature, respectively. When the heat flux is not very high, the Eq. (2.53) can be transformed into a governing equation of T_e:

$$\begin{gathered}\tau_e \rho_e C_e \frac{\partial^2 T_e}{\partial t^2} + \rho_e C_e \frac{\partial T_e}{\partial t} = \kappa_e \left(1 - \frac{\rho_e C_e \tau_e u_h^2}{\kappa_e} \right) \frac{\partial^2 T_e}{\partial x^2} - 2\tau_e u_h \rho_e C_e \frac{\partial^2 T_e}{\partial t \partial x} + \\ \left\{ S - G(T_e - T_l) + \tau_e \frac{\partial}{\partial t} [S - G(T_e - T_l)] + 2\tau_e u_h \frac{\partial}{\partial x} [S - G(T_e - T_l)] \right\}\end{gathered} \tag{2.54}$$

where ρ_e, C_e and κ_e are the density, specific heat and thermal conductivity of electrons, $\tau_e = \frac{2\rho_e^2 C_e \kappa_e}{5\pi^2 n^3 k_B^3 T}$ is the characteristic time of electrons.

For lattice system, the mass and momentum conservation equations are listed as:

$$\frac{\partial \rho_h}{\partial t} + \rho_h \frac{\partial u_h}{\partial x} + u_h \frac{\partial \rho_h}{\partial x} = \frac{G(T_e - T_l)}{c^2} \tag{2.55}$$

$$\rho_h \left(\frac{\partial u_h}{\partial t} + u_h \frac{\partial u_h}{\partial x} \right) + \frac{u_h}{c^2} \left[G(T_e - T_l) \right] + \frac{\partial P_h}{\partial x} + f_h = 0 \tag{2.56}$$

The governing equation of T_l is given as:

$$\begin{gathered} \tau_l \rho_l C_l \frac{\partial^2 T_l}{\partial t^2} + \rho_l C_l \frac{\partial T_l}{\partial t} = \kappa_l \left(1 - \frac{\rho_l C_l \tau_l u_h^2}{\kappa_l} \right) \frac{\partial^2 T_l}{\partial x^2} - 2\tau_l u_h \rho_l C_l \frac{\partial^2 T_l}{\partial t \partial x} + \\ \left\{ G(T_e - T_l) + \tau_l \frac{\partial}{\partial t} \left[G(T_e - T_l) \right] + 2\tau_l u_h \frac{\partial}{\partial x} \left[G(T_e - T_l) \right] \right\} \end{gathered} \tag{2.57}$$

where ρ_l, C_l and κ_l are the density, specific heat, and thermal conductivity of lattice, $\tau_l = \frac{\kappa_l}{2\gamma \rho_l C_l^2 T}$ is the characteristic time of lattice.

It is noted that the Eqs. (2.54) and (2.57) are both damped wave equations, where the damping terms are decided by different characteristic times. τ_e is about 10^{-15} s, much shorter than τ_l of 10^{-12} s, thus the thermalization process of electrons happens very fast while T_l is almost a constant. The thermal wave propagation in metals can be quantitatively studied using the Eqs. (2.54) and (2.57), if the characteristic times equal zero, these equations will reduce to heat diffusion equations.

2.2.4 Numerical Simulation Examples

Some numerical simulation examples of TSTM, C–V model, hyperbolic two-step (HTS) model, parabolic two-step (PTS) model, and dual-phase lag (DPL) model are given for comparison. All these models are solved using a finite-difference scheme in double precision. The material properties of Au film used in calculation are: $\kappa = 315\,\mathrm{Wm^{-1}K^{-1}}$, $\rho_l C_l = 2.5 \times 10^6\,\mathrm{Jm^{-3}K^{-1}}$, $G = 2.6 \times 10^{16}\,\mathrm{Wm^{-3}K^{-1}}$, $C_e = 70 T_e\,\mathrm{Jm^{-3}K^{-1}}$. The numerical results are shown in the Figs. 2.6 and 2.7.

The PTS model has a parabolic equation, so no temperature fluctuation can be observed in the calculated temperature curve. The other four models have hyperbolic wave equations, which are capable to predict the propagation of thermal waves. A typical expression of damped wave equation is given as:

$$\frac{\partial^2 T}{\partial t^2} + A \frac{\partial T}{\partial t} = B \frac{\partial^2 T}{\partial x^2} + C \tag{2.58}$$

Different A, B, and C coefficients are responsible for different models:

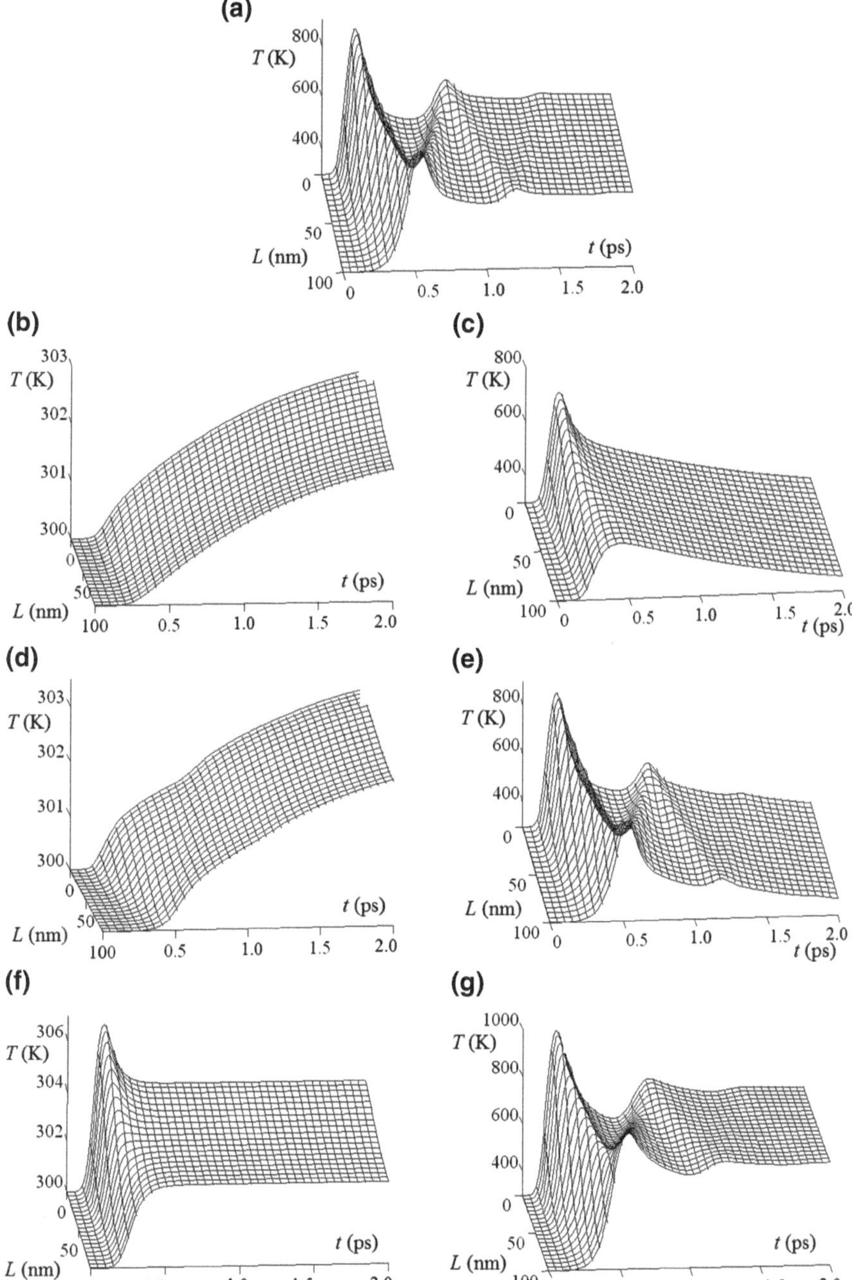

Fig. 2.6 Electron and lattice temperature *curves* of different models 1. **a** C–V model electron temperature, **b** PTS model lattice temperature, **c** PTS model electron temperature, **d** HTS model lattice temperature, **e** HTS model electron temperature, **f** DPL model lattice temperature, **g** DPL model electron temperature

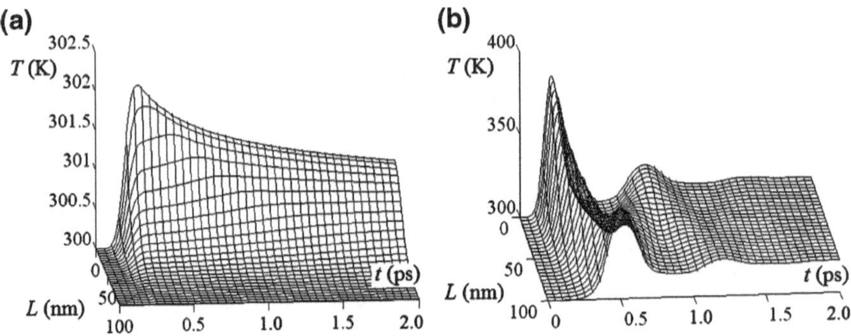

Fig. 2.7 Electron and lattice temperature curves of different models 2. **a** DPL model lattice temperature, **b** DPL model electron temperature

Table 2.1 Coefficient selection for different models

Models	C–V	PTS	HTS
A	$\frac{1}{\tau_0}$	$G\left(\frac{1}{\rho_e C_e}+\frac{1}{\rho_l C_l}\right)$	$\frac{1}{\tau_e}+\frac{G}{\rho_e C_e}$
B	$\frac{\kappa}{\rho C \tau_0}$	$\frac{\kappa_e G}{\rho_e C_e \rho_l C_l}$	$\frac{\kappa_e}{\rho_e C_e \tau_e}$
C	$\frac{1}{\rho C \tau_0}\left(S+\tau_0\frac{\partial S}{\partial t}\right)$	$\frac{G}{\rho_e C_e \rho_l C_l}\left(S+\frac{\rho_l C_l}{G}\frac{\partial S}{\partial t}\right)+\frac{\kappa_e}{\rho_e C_e}\frac{\partial^3 T_e}{\partial t \partial x^2}$	$\frac{1}{\rho_e C_e \tau_e}\left(S+\tau_e\frac{\partial S}{\partial t}\right)+\frac{G}{\rho_e C_e}\left(\frac{\partial T_l}{\partial t}-\frac{T_e-T_l}{\tau_e}\right)$

Models	DPL	TSTM
A	$\frac{1}{\tau_q}$	$\frac{1}{\tau_e}$
B	$\frac{\kappa}{(\rho_e C_e+\rho_l C_l)\tau_q}$	$\frac{\kappa}{\rho C \tau_e}-u_h^2$
C	$\frac{1}{(\rho_e C_e+\rho_l C_l)\tau_q}\left(S+\tau_q\frac{\partial S}{\partial t}\right)+\frac{\kappa \tau_T}{(\rho_e C_e+\rho_l C_l)\tau_q}\frac{\partial^3 T}{\partial x^2 \partial t}$	$\frac{1}{\rho C \tau_e}\left(S+\tau_e\frac{\partial S}{\partial t}+2\tau_e u_h\frac{\partial S}{\partial x}\right)-2u_h\frac{\partial^2 T}{\partial t \partial x}$

From the Table 2.1, one can conclude that:

1. The coefficient of diffusion term A decreases as the relaxation time increases for hyperbolic models, while it increases as the coupling factor G increases in the PTS model;
2. At the same relaxation time, the propagation speed of thermal wave increases as the thermal conductivity increases, decreases as the specific heat increases;

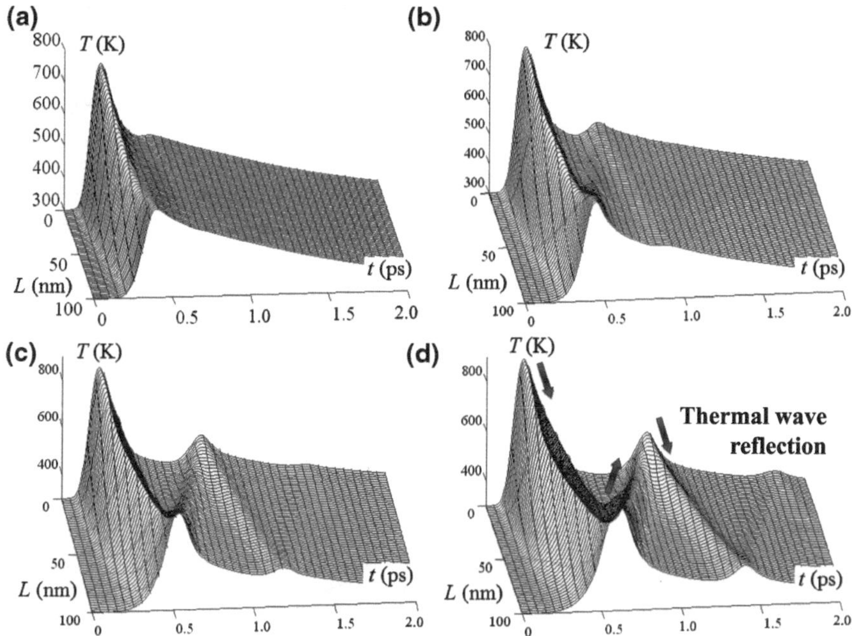

Fig. 2.8 Electron temperature curves predicted by the HTS model at different relaxation times. **a** $\tau_0 = 0.04\,\text{ps}$, **b** $\tau_0 = 0.08\,\text{ps}$, **c** $\tau_0 = 0.16\,\text{ps}$, **d** $\tau_0 = 0.24\,\text{ps}$

3. Under the same internal heat source condition, the higher specific heat results in smaller temperature responses to the input thermal disturbances, the thermal wave phenomena become less obvious;
4. By giving proper characteristic times, the TSTM model can be reduced to the other kinds of models.

Thermal wave is understood as the result of non-negligible TM inertia. In TM theory, the characteristic time represents the inertia effect of thermon gas. The longer characteristic time, the stronger TM inertia will be. Meanwhile, the higher specific heat of material has more ability to resist the temperature change. So the longer characteristic time and smaller specific heat will cause more obvious thermal wave phenomena. Because the specific heat of lattices is 100 times larger than that of electrons, no temperature wave propagation is observed for lattice temperature. The characteristic time equals zeros for the PTS model, thus no thermal wave is observed either.

Different relaxation times 0.04, 0.08, 0.16, and 0.24 ps are used for numerical calculation in the Fig. 2.8.

It is seen in the Fig. 2.8 that a longer relaxation time results in more obvious thermal wave behaviors. The thermal wave can be reflected at the surfaces of the film as shown in the Fig. 2.8d, the arrows show the propagation directions of the thermal wave.

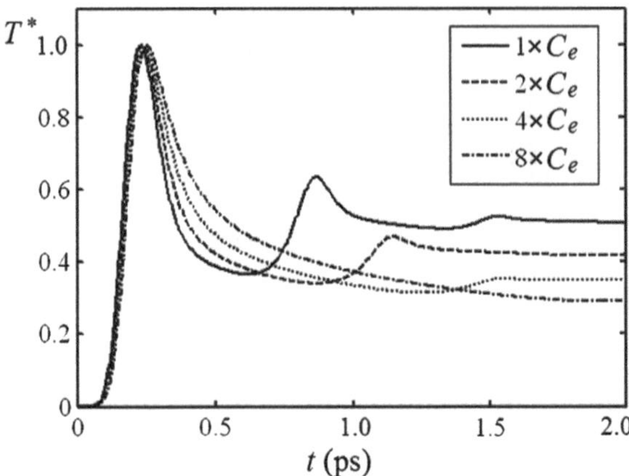

Fig. 2.9 Normalized electron temperature predicted by the C–V model at different specific heats

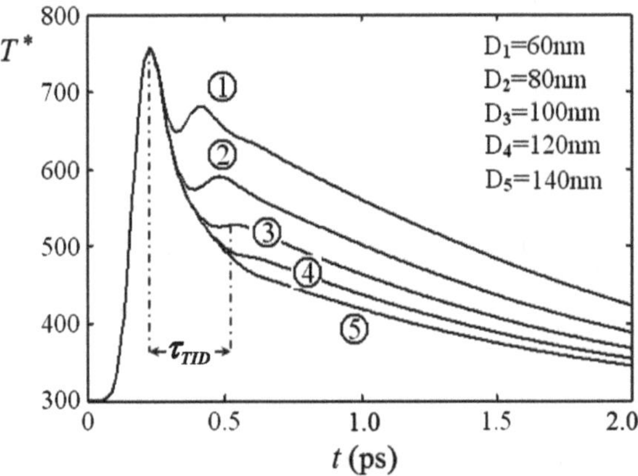

Fig. 2.10 Electron temperature profiles of different film thicknesses when $\tau = 0.4\,\text{ps}$

It is seen in the Fig. 2.9 that the temperature fluctuation is weakened as the specific heat increases, consistent with the theoretical analysis above. When $8C_e$ is used, the thermal wave disappears and the heat diffusion dominates.

It is also noted that the thermal wave will attenuate during its propagation, similar to the mechanical wave and electromagnetic wave. In order to describe the attenuation of thermal wave quantitatively, a thermal wave time influence domain τ_{TID} has been defined. τ_{TID} is the time needed for the thermal wave to disappear in its direction of propagation, long τ_{TID} means that the attenuation effect is weak and the thermal wave has a larger influence domain. Figures 2.9 and 2.10 show the propagation of thermal wave inside the Au film. For thin films, the thermal wave will be reflected from the rear surface and cause a second temperature peak at the front surface. As the film thickness increases, the thermal wave will be attenuated and the second temperature peak disappears. The time interval between two temperature peaks is defined as τ_{TID}.

Five film thicknesses from 60 to 140 nm are used for calculation as shown in the Fig. 2.10.

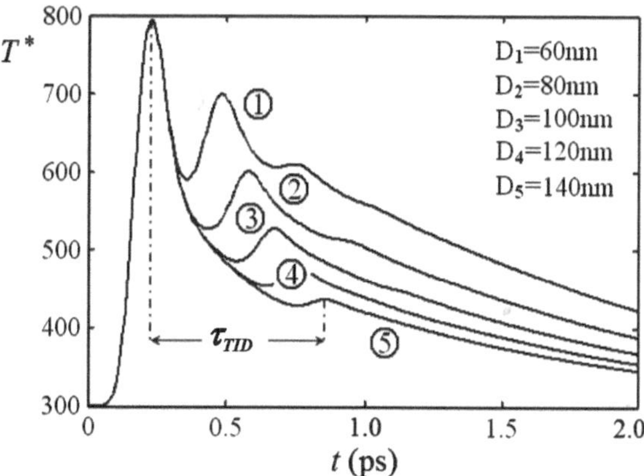

Fig. 2.11 Electron temperature profiles of different film thicknesses when $\tau = 0.8$ ps

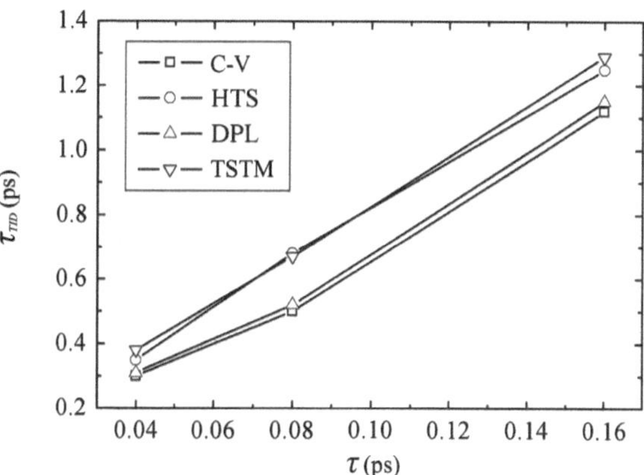

Fig. 2.12 Thermal wave time influence domain plotted with respect to the relaxation time

Based on the calculation results of Figs. 2.10 and 2.11, $\tau_{\mathrm{TID}} = 0.55 - 0.20 = 0.35$ ps when $\tau = 0.4$ ps; $\tau_{\mathrm{TID}} = 0.88 - 0.20 = 0.68$ ps when $\tau = 0.8$ ps. It is seen that the influence domain of thermal wave increases as the relaxation time increases. The calculation results of different models are summarized in the Fig. 2.12.

Figure 2.12 compares the thermal wave time influence domains of different models. At the same relaxation time, the HTS and TSTM models predict larger τ_{TID} than the DPL and C–V models.

The TSTM model is able to avoid some non-physical problems caused by other thermal wave models. For example, the other thermal wave models will cause negative temperatures during the thermal wave propagation at very low temperatures, while the temperature predicted by the TSTM model keeps positive.

Figure 2.13 shows the physical model of a heat conductor inside which two thermal waves propagates in the opposite directions. At a certain time, these two thermal waves will meet each other in the middle, several thermal wave models are solved numerically to predict the temperature responses. Besides the thermal wave model,

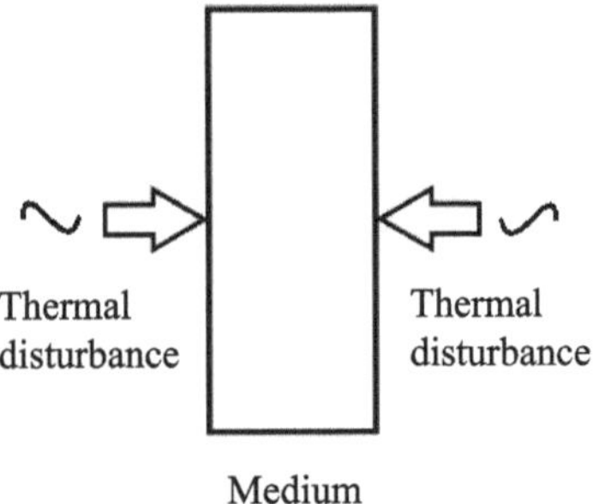

Fig. 2.13 Schematic diagram of two thermal waves propagating in the opposite directions

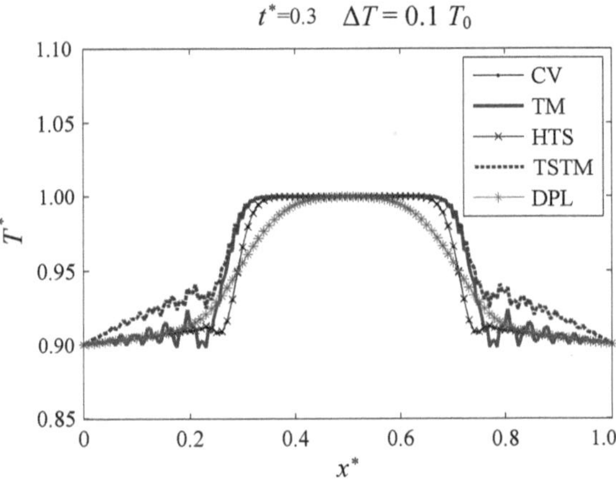

Fig. 2.14 Temperature distribution under the small thermal disturbance condition when $t^* = 0.3$

the other models will cause negative temperature which is against the third law of thermodynamics [4].

Here, T^*, x^* and t^* are the normalized temperature, length, and time, respectively. T is the environment temperature. When $t^* = 0$, the boundary temperature was suddenly decreased to $0.9T$ and two thermal waves started to propagate in the opposite directions. After a while, these two thermal waves meet each other in the middle position. The Fig. 2.14 gives the predicted temperature responses when $t^* = 0.3$ The TM model is for dielectrics and the TSTM model is for metals. The C–V model is solved using the properties of electron gas and the calculation result is consistent with that of the HTS model.

The Fig. 2.15 gives the calculation result when $t^* = 0.9$. It is seen that temperature in the middle is decreased from the initial temperature T when two thermal waves are superimposed. This can be seen as the interference effect of thermal waves, similar to other kinds of waves, such as light, radio, acoustic, and surface water waves. If the initial temperature T is close to zero and the fluctuated temperature amplitude is large, the negative temperature may occur.

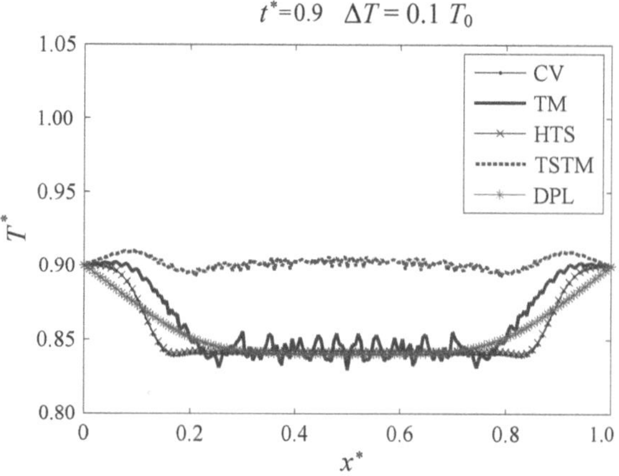

Fig. 2.15 Temperature distribution under the small thermal disturbance condition when $t^* = 0.9$

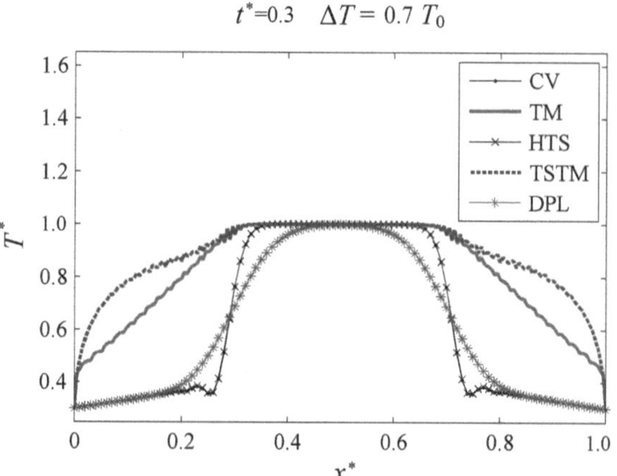

Fig. 2.16 Temperature distribution under the large thermal disturbance condition when $t^* = 0.3$

The Figs. 2.16 and 2.17 show the calculation results when $t^* = 0.3$ and $t^* = 0.9$, where the initial temperature was decreased to $0.3T$. In this case, the initial temperature change ΔT is increased from 0.1 to $0.7T$, the superimposed thermal waves cause negative temperatures in the C–V, HTS, and DPL models, while the TM and TSTM models predict positive temperatures all along. This is because the other models give thermal wave equations by introducing constant relaxation times to the heat diffusion equation, lacking the deep understanding of the physical essence of the relaxation time; in this case, the thermal wave equation has the same expression as the other mechanical wave equations. For mechanical waves, it is correct to observe the negative mechanical wave amplitudes (below the initial equilibrium state); but for thermal waves, the mechanical wave equation is not applicable when T is close to zero.

In the TM model, the thermal wave is known to be caused by the non-negligible inertia effect between the flow of TM and its driven force (temperature gradient).

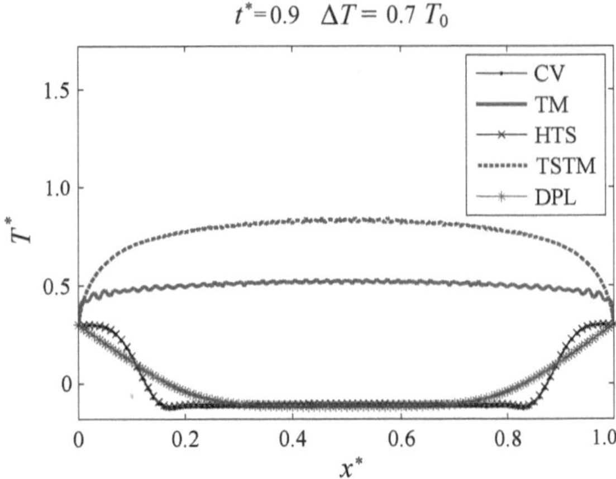

Fig. 2.17 Temperature distribution under the large thermal disturbance condition when $t^* = 0.9$

Further a general heat conduction equation has been developed based on the first principles, in which the characteristic time is no longer simply a constant, but decided by the material properties, temperature, microscopic distribution function, etc. The general equation is no longer a simple mechanical wave equation, but related with the heat flux, temporal, and spatial partial derivatives of temperature, etc. As a result, the general heat conduction equation is able to predict the thermal wave behavior correctly without any paradoxes.

2.3 Non-Fourier Heat Conduction Equation in Steady States

In steady states, all the other thermal wave models will reduce to Fourier's law, implying no non-Fourier heat conduction existed. But the general heat conduction equation remains different from the Fourier's equation in steady states, giving the first theoretical prediction for the steady non-Fourier heat conduction. One-dimensional steady general heat conduction equation is given as:

$$\left(\kappa - \tau_h \frac{q^2}{\rho C_v T^2}\right)\frac{\partial^2 T}{\partial x^2} + \frac{2\tau_h q}{\rho C_v T}\frac{\partial S}{\partial x} + S = 0 \tag{2.59}$$

For a constant S, the Eq. (2.59) can be simplified as:

$$\left(\kappa - \tau_h \frac{q^2}{\rho C_v T^2}\right)\frac{\partial^2 T}{\partial x^2} + S = 0 \tag{2.60}$$

Compare with the Fourier's heat diffusion equation, the non-Fourier behavior is mainly caused by the first term in parentheses, which is referred to as the spatial

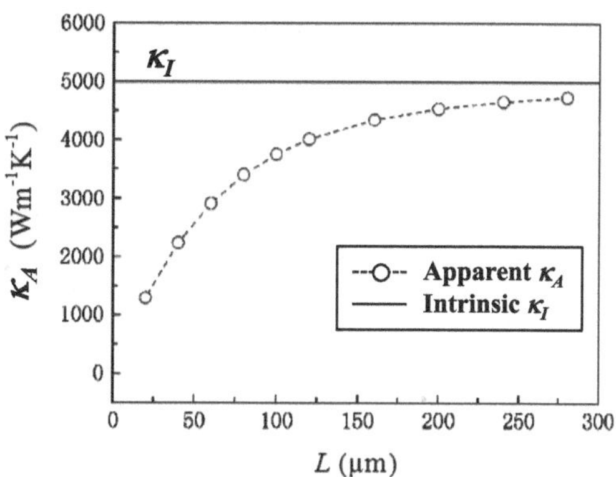

Fig. 2.18 Thermal conductivity of carbon nanotube plotted with respect to length[12]. Reprinted with the permission from Ref. [12]. Copyright [2007], AIP Publishing LLC

inertia effect of TM. Two kinds of thermal conductivities can be defined from the Eq. (2.60), which are the apparent thermal conductivity κ_A and the intrinsic thermal conductivity κ_I:

$$\kappa_A = \kappa_I - \tau_h \frac{q^2}{\rho C_v T^2} = \kappa_I \left(1 - \frac{q^2}{2\gamma_h \rho^2 C_v^3 T^3}\right) \tag{2.61}$$

The Eq. (2.60) can be also written as:

$$\kappa_A \frac{\partial^2 T}{\partial x^2} + S = 0 \tag{2.62}$$

To avoid confusion, in this thesis, κ_A is referred to as the apparent thermal conductivity, κ_I or κ are referred to as the intrinsic thermal conductivity. The Eq. (2.62) implies that the ratio between heat flux and temperature gradient is the apparent thermal conductivity. κ_A equals κ_I under the common conditions, but under the ultra-high heat flux conditions or at very low temperatures, κ_A will be less than κ_I. Strictly speaking, κ_A is not one of the material properties, because the material property should be decided only by crystal structures, independent with the process properties, such as heat flux, electrical current, size, etc.

The difference between κ_A and κ_I is shown in the Fig. 2.18 in Ref. [12].

The temperature difference between two ends of carbon nanotube (CNT) is controlled to be 20 K in the calculation and the average temperature is 70 K. As the length of CNT increases, the temperature gradient and heat flux decrease. Thus κ_A increases according to the Eq. (2.61). The Fig. 2.18 shows the calculation result that κ_A approaches κ_I as the length increases. The length-dependent thermal conductivity of CNT has been reported in both experiment and molecular simulation [21–24]. The TM theory points out that the spatial inertia effect of TM causes the length-dependent thermal conductivity and gives a prediction from first principles.

2.4 Heat Flow Choking Phenomenon

It is noted from the Eq. (2.60) that a negative thermal conductivity may occur if $\frac{q^2}{2\gamma_h \rho^2 C_v^3 T^3} > 1$. This is against the physical reality. As a matter of fact, a new physical phenomenon "Heat flow choking" is predicted based on this observation. Let us take a look at the choked flow effect: when a subsonic flowing fluid passes through a restriction, the fluid velocity, and therefore the Mach number, increases as the pressure increases for a given environment pressure. The choked flow occurs as the limiting condition when the Mach number equals unity, in this case, the mass flow rate stops to increase with a further increase in the upstream pressure. Similarly, the thermon gas is also compressible and the heat flow choking occurs when the Mach number of the thermon gas equals unity. The propagation speed of thermal disturbance (thermal sound speed) is given from the Eq. (2.51):

$$C_h = \sqrt{\frac{\kappa_I}{\rho C_v \tau_h}} \tag{2.63}$$

For dielectrics, the characteristic time $\tau_h = \frac{\kappa}{2\gamma_h \rho C_v^2 T}$, thus the Eq. (2.63) can be simplified as:

$$C_h = \sqrt{2\gamma_h C_v T} \tag{2.64}$$

It is noted that the Eq. (2.64) is very similar with the expression of sound speed $v_s = [r(r-1)C_v T]^{1/2}$. Further the thermal Mach number Ma_h is defined as the ratio between the drift velocity of thermon gas u_h and the thermal sound speed C_h:

$$\mathrm{Ma}_h = \frac{u_h}{C_h} \tag{2.65}$$

In steady states, the general heat conduction Eq. (2.60) can be simplified as:

$$\kappa_I \left(1 - \mathrm{Ma}_h^2\right) \frac{\partial^2 T}{\partial x^2} + S = 0 \tag{2.66}$$

As mentioned above, $\mathrm{Ma}_h = 1$ is the limiting condition for the heat flow choking. The flows of thermon gas and ideal gas are compared in the Fig. 2.19.

The Fig. 2.19 shows a schematic diagram of thermon gas flow in a heat conductor. Just like the flow of compressible air in a convergent nuzzle, the velocity increases and the pressure decreases in the flow direction. The drift velocity of thermon gas increases as it flows in the opposite direction of temperature gradient, meanwhile, the pressure of thermon gas or saying temperature, decreases as well. The pressure change of air flow and temperature change of thermon gas flow are shown in the Fig. 2.20.

As shown in the Fig. 2.20, Mach number equaling unity is the criterion condition for flow choking. In the air flow, a pressure jump occurs at the exit of the nozzle

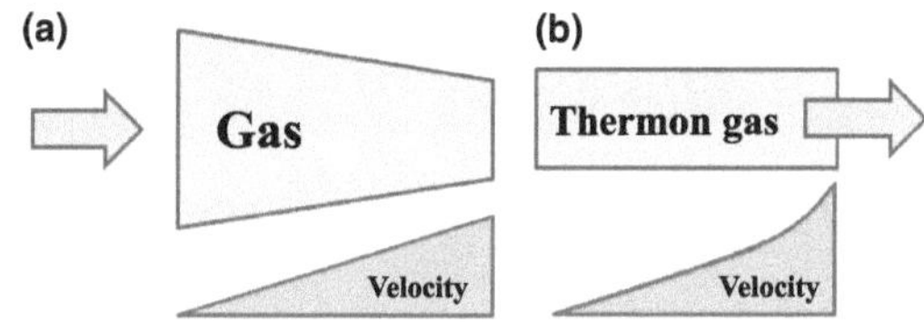

Fig. 2.19 Comparison between the flows of thermon gas and ideal gas. **a** Flow of gas in a contraction nozzle. **b** Flow of thermon gas in a conductor with equal area. Reprinted from "Heat flow choking in carbon nanotubes, 53, Hai-Dong Wang, Bing-Yang Cao, Zeng-Yuan Guo, 1796–1800". Copyright [2010], with permission from Elsevier

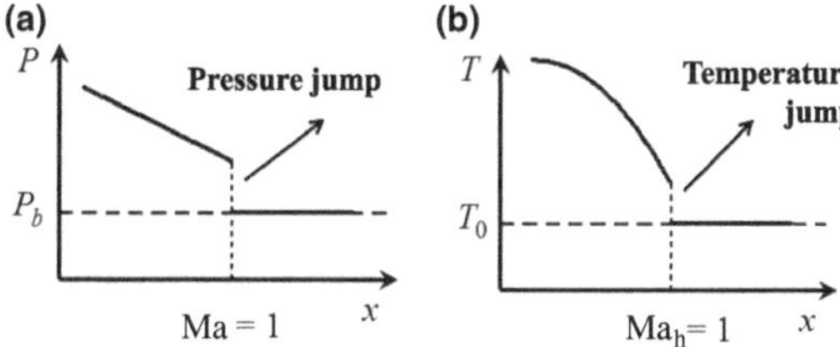

Fig. 2.20 Comparison between the *pressure jump* and *temperature jump* induced by air flow choking and heat flow choking. **a** Pressure profile along the flow direction. **b** Temperature profile along the flow direction. Reprinted from "Heat flow choking in carbon nanotubes, 53, Hai-Dong Wang, Bing-Yang Cao, Zeng-Yuan Guo, 1796–1800". Copyright [2010], with permission from Elsevier

(throat position) when Ma $= 1$; meanwhile in the thermon gas flow, a temperature jump occurs at the cold side of the heat conductor when $\mathrm{Ma}_h = 1$. The velocity gradients of air flow and thermon gas flow are given as:

$$\textit{Airflow}: \frac{du}{dx} = -\frac{1}{\rho u}\frac{dP}{dx} \tag{2.67}$$

$$\textit{Thermongasflow}: \frac{du_h}{dx} = \frac{S}{\rho C_v T} - \frac{q}{\rho C_v T^2}\frac{dT}{dx} \tag{2.68}$$

The first term on the right side of the equal sign in the Eq. (2.68) is caused by the internal heat source. The flows of air and thermon gas are driven by the pressure gradient and temperature gradient with different coefficients, the latter coefficient is proportional to the reciprocal of temperature square. A critical temperature for heat flow choking can be extracted from the equation $\mathrm{Ma}_h = 1$ as:

$$T_c = \left(\frac{q^2}{2\gamma_h \rho^2 C_v^3}\right)^{1/3} \tag{2.69}$$

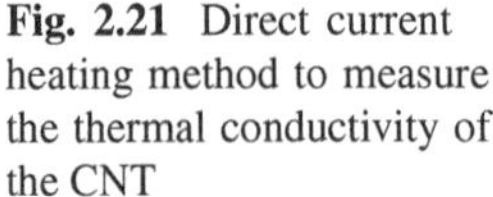

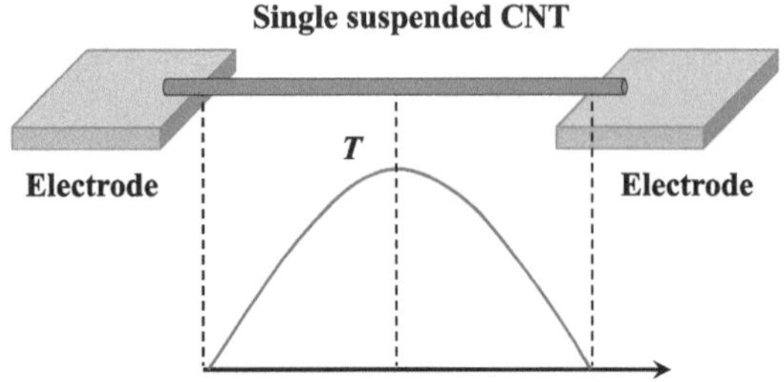

Fig. 2.21 Direct current heating method to measure the thermal conductivity of the CNT

For given material properties, T_c increases as the heat flux q increases, the temperature jump increases as well. Normally, T_c is much smaller than the room temperature and the general heat conduction law can be reduced to Fourier's law. Only for ultra-high heat flux or at very low temperatures, the heat flow choking can be observed in the experiment. The heat flow choking is caused by the spatial TM inertia effect, which was noticed as "thermal inertia" in relativistic thermohydrodynamics [25].

The heat flux in CNT can be ultra-high because of its small cross-sectional area and high burning temperature, making the CNT an ideal material to observe heat flow choking. The Fig. 2.21 gives a schematic diagram of the experiment measuring the thermal conductivity of CNT.

In the Fig. 2.21, an individual CNT is suspended above the substrate with two ends connected with metallic electrodes. In the experiment, a direct current is used to heat the CNT and a reversed parabolic temperature distribution is established along the nanotube. Joule heat generated in the CNT flows from the middle point to the two ends, the heat flux increases as the electrical power increases. When Ma_h equals unity, the heat flow choking can be observed. In this case, the critical heat flux is $q_c = \sqrt{2\gamma\rho^2 C^3 T_0^3}$ and the temperature jump at the end is $\Delta T = \sqrt[3]{q^2/(2\gamma\rho^2 C^3)} - T_0$. The general conduction law can be used to predict the temperature of the CNT. The material properties are chosen as: $\rho = 1,400\,\mathrm{kgm^{-3}}$, $C = 500\,\mathrm{Jkg^{-1}K^{-1}}$, $D = 1.8\,\mathrm{nm}$, $L = 10\,\mathrm{\mu m}$, $\gamma = 1$, $T_0 = 300\,\mathrm{K}$, $S = 1\,\mathrm{\mu W}$ and $q_c = 1.15 \times 10^{11}\,\mathrm{Wm^{-2}}$. The cross-sectional area of the CNT is calculated as $A = \pi D d$, where the shell thickness is $d = 0.34\,\mathrm{nm}$. A finite-difference scheme with double precision is used for numerical calculations.

In the Fig. 2.22, $x = 0$ is the middle position of CNT and the length is $10\,\mathrm{\mu m}$. The heat flux and drift velocity increase in the x direction, while the thermal sound speed decreases. The heat flow choking occurs at the right end of CNT when $\mathrm{Ma}_h = 1$ as shown in the inset of the Fig. 2.22. The drift velocity reaches $700\,\mathrm{ms^{-1}}$ at the right end. The calculation results with different intrinsic thermal conductivities are compared in the figure, the higher thermal conductivity results in higher drift velocity, but the temperature jump at the end remains the same.

The maximum heat flux in the calculation is $2.60 \times 10^{11}\,\mathrm{Wm^{-2}}$, which is higher than the critical heat flux, thus a temperature jump of 200 K occurs at the end of CNT as shown in the Fig. 2.23. Because of the temperature jump, the temperature predicted by the general heat conduction law is notably higher than that predicted by

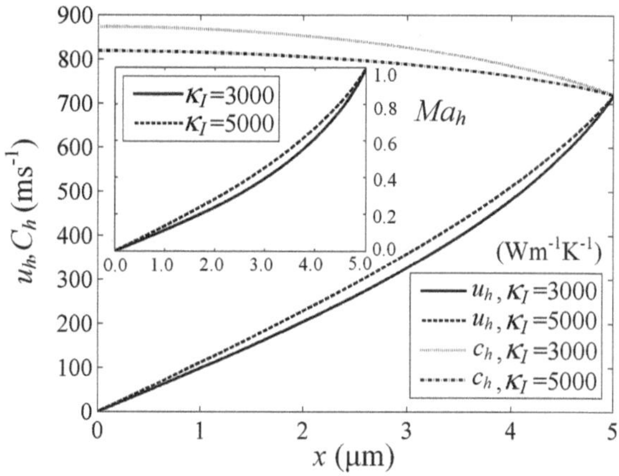

Fig. 2.22 Drift velocity and thermal sound speed of CNT with different intrinsic thermal conductivities. Reprinted from "Heat flow choking in carbon nanotubes, 53, Hai-Dong Wang, Bing-Yang Cao, Zeng-Yuan Guo, 1796–1800". Copyright [2010], with permission from Elsevier

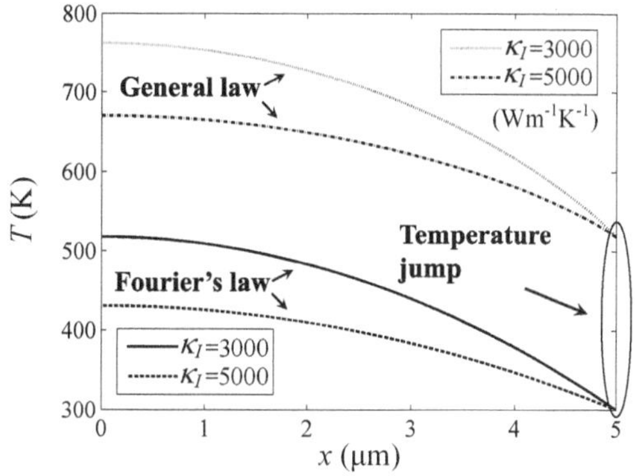

Fig. 2.23 Temperature distributions along the CNT predicted by the general heat conduction law and Fourier's law. Reprinted from "Heat flow choking in carbon nanotubes, 53, Hai-Dong Wang, Bing-Yang Cao, Zeng-Yuan Guo, 1796–1800". Copyright [2010], with permission from Elsevier

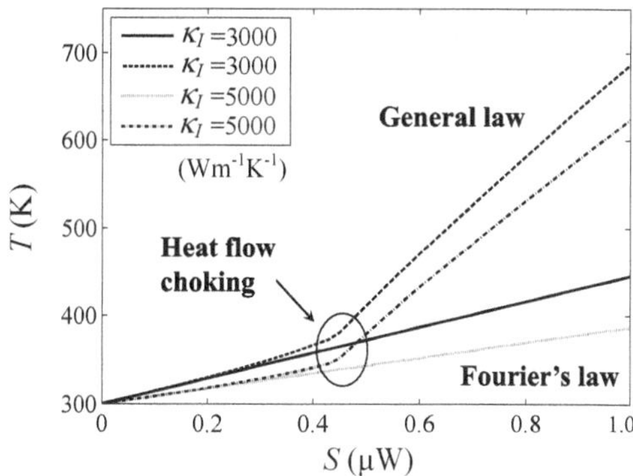

Fig. 2.24 Average temperature of CNT varied with the electrical power. Reprinted from "Heat flow choking in carbon nanotubes, 53, Hai-Dong Wang, Bing-Yang Cao, Zeng-Yuan Guo, 1796-1800". Copyright [2010], with permission from Elsevier

Fourier's law. Before the proposition of TM theory, the thermal conductivity of CNT is extracted by Fourier's law only, in this way, the reported thermal conductivities are all apparent ones that are related with the heat flux.

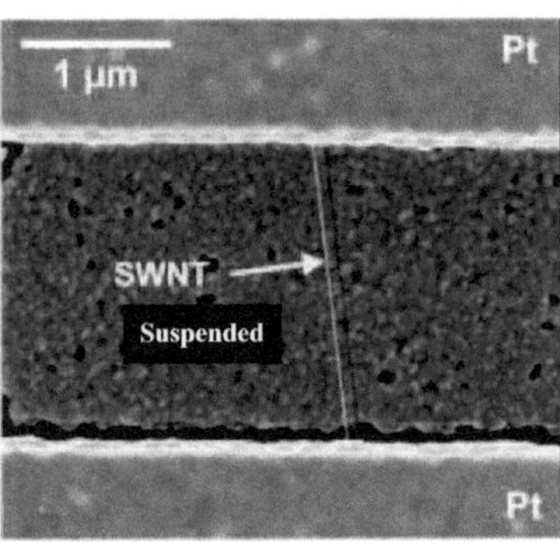

Fig. 2.25 SEM image of the prepared SWCNT sample [26]. Reprinted with the permission from Ref. [26]. Copyright [2006], American Chemical Society

The Fig. 2.24 shows the average temperature of CNT at different electrical powers. Both temperatures predicted by Fourier's law and the general heat conduction law are plotted for comparison. For a constant thermal conductivity, the temperature based on Fourier's law is proportional to the electrical power. Meanwhile, the temperature curve predicted by the general law has a turning point at 0.44 μW, which is the critical electrical power for appearance of the heat flow choking. Before the turning point, the temperature predicted by the general law is almost the same as that predicted by Fourier's law; after the turning point, the previous temperature increases significantly higher than the latter one because of the increasing temperature jump at the end. Once the heat flow choking occurs, the temperature of CNT could be notably underestimated by Fourier's law, this may cause thermal failure of the CNT-based devices in practical applications.

The experimental result of Pop [26] has been studied to show the effect of heat flow choking on heat conduction process. A single-walled carbon nanotube (SWCNT) with 1.7 nm in diameter and 2.6 μm in length has been prepared, and then a direct current heating method was used to measure the thermal conductivity of SWCNT. A scanning electron microscope (SEM) image of the SWCNT is shown in the Fig. 2.25 from the Ref. [26].

As shown in the Fig. 2.25, an individual SWCNT is suspended between two platinum (Pt) electrodes for measurement. The Ref. [26] gives a measured current–voltage curve of SWCNT as shown in the Fig. 2.26.

The average temperature of SWCNT can be extracted from the precisely measured resistance using Landauer–Büttiker model [27, 28]:

$$R(V,T) = R_c + \frac{h}{4q_e^2}\frac{L + \lambda_{\text{eff}}(V,T)}{\lambda_{\text{eff}}(V,T)} \tag{2.70}$$

$$\lambda_{\text{eff}} = \left(\lambda_{AC}^{-1} + \lambda_{\text{OP,ems}}^{-1} + \lambda_{\text{OP,abs}}^{-1}\right)^{-1} \tag{2.71}$$

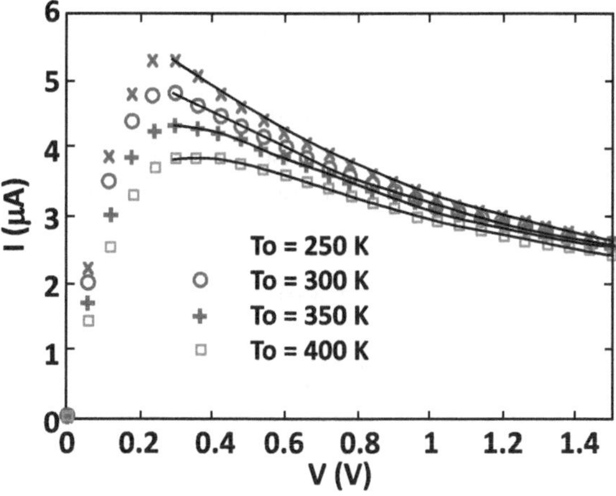

Fig. 2.26 Current–voltage curve of an individual SWCNT [26]. Reprinted with the permission from Ref. [26]. Copyright [2006], American Chemical Society

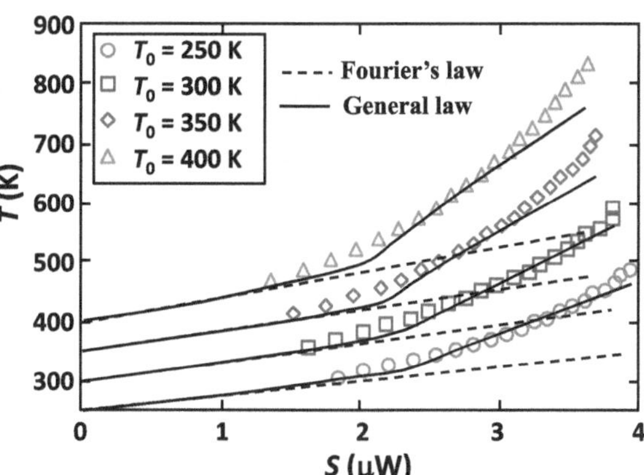

Fig. 2.27 Average temperature of SWCNT plotted with respect to the electric power. Reprinted from "Heat flow choking in carbon nanotubes, 53, Hai-Dong Wang, Bing-Yang Cao, Zeng-Yuan Guo, 1796–1800". Copyright [2010], with permission from Elsevier

where h, q_e and λ_{eff} are the Planck constant, electric charge element and effective electron mean free path, respectively. $h/(4q_e^2) = 6.5\,\text{K}\Omega$ is the quantum contact resistance. Based on the Eqs. (2.70) and (2.71), the temperature–electrical power curve of SWCNT can be plotted in the Fig. 2.27:

The average temperatures predicted by Fourier's law and the general heat conduction law are also plotted in the Fig. 2.27 to compare with the experimental data. Here, a constant thermal conductivity is used for theoretical calculation. It is seen that the general heat conduction law matches well with the experimental data while Fourier's law gives significantly lower predictions. Under the experimental conditions, the drift velocity of thermon gas in the SWCNT is about several hundreds of meters per second, in this case, the assumption that the TM resistance is proportional to the drift velocity is not valid and a second-order correction to the TM resistance is needed.

$$\frac{\partial \rho_h}{\partial t} + \rho_h \frac{\partial u_h}{\partial x} + u_h \frac{\partial \rho_h}{\partial x} = \frac{S}{c^2} \tag{2.72}$$

$$\rho_h\left(\frac{\partial u_h}{\partial t}+u_h\frac{\partial u_h}{\partial x}\right)+u_h\frac{S}{c^2}+\frac{\partial P_h}{\partial x}+f_h=0 \tag{2.73}$$

The Eqs. (2.72) and (2.73) are the governing equations for the motion of thermon gas, based on which, an one-dimensional steady expression of the general heat conduction equation with internal heat source S can be derived as:

$$\begin{aligned}&\left(\frac{-\kappa_I qS}{\gamma\rho^2C^3T^3}\frac{\partial T}{\partial x}+\frac{\kappa_I q^2}{2\gamma\rho^2C^3T^4}\left(\frac{\partial T}{\partial x}\right)^2\right)+\frac{\kappa_I}{\gamma\rho C^2T}\left(\frac{S}{\rho CT}-\frac{q}{\rho CT^2}\frac{\partial T}{\partial x}\right)\\&\times\left(S-\frac{q}{T}\frac{\partial T}{\partial x}\right)+\kappa_I\left(1-\frac{q^2}{2\gamma\rho^2C^3T^3}\right)\frac{\partial^2 T}{\partial x^2}+S=0\end{aligned} \tag{2.74}$$

In order to simplify the Eq. (2.74), an approximate integration equation is given as:

$$q\approx-\kappa_I\frac{\partial T}{\partial x}\approx\frac{1}{2}SL \tag{2.75}$$

Substituting the Eq. (2.75) into the Eq. (2.74), one can get:

$$\kappa_I\left(1-\frac{q^2}{2\gamma\rho^2C^3T^3}\right)\frac{d^2T}{dx^2}+\left(1+\frac{\varepsilon\kappa_I S}{\gamma\rho^2C^3T^2}\right)S\approx0 \tag{2.76}$$

where the parameter ε is:

$$\varepsilon=\frac{\beta_S^*}{4}+\frac{\beta_S^{*2}}{32}+\left(1+\frac{\beta_S^*}{4}\right)^2 \tag{2.77}$$

where the normalized parameter for the internal heat source is: $\beta_S^*=\frac{SL^2}{T\kappa_I}$, L is the length. A normalized TM resistance is given as:

$$F_h=\beta_1^*U_h+\frac{1}{2}\beta_2^*U_h^2+\frac{1}{6}\beta_3^*U_h^3+\ldots \tag{2.78}$$

where $F_h=\frac{L}{P_{h0}}f_h$ and $P_{h0}=\frac{\gamma\rho C^2T_0^2}{c^2}$. Ignoring the high order small quantities, a second TM resistance can be written as:

$$f_h\approx\frac{P_{h0}}{L}\left(\beta_1^*U_h+\frac{1}{2}\beta_2^*U_h^2\right)=\beta_1u_h+\beta_2u_h^2 \tag{2.79}$$

where the first-order coefficient is:

$$\beta_1=\frac{2\gamma\rho^2C^3T^2}{\kappa_I c^2}\quad and\quad \beta_1^*=\frac{T^2}{T_0^2} \tag{2.80}$$

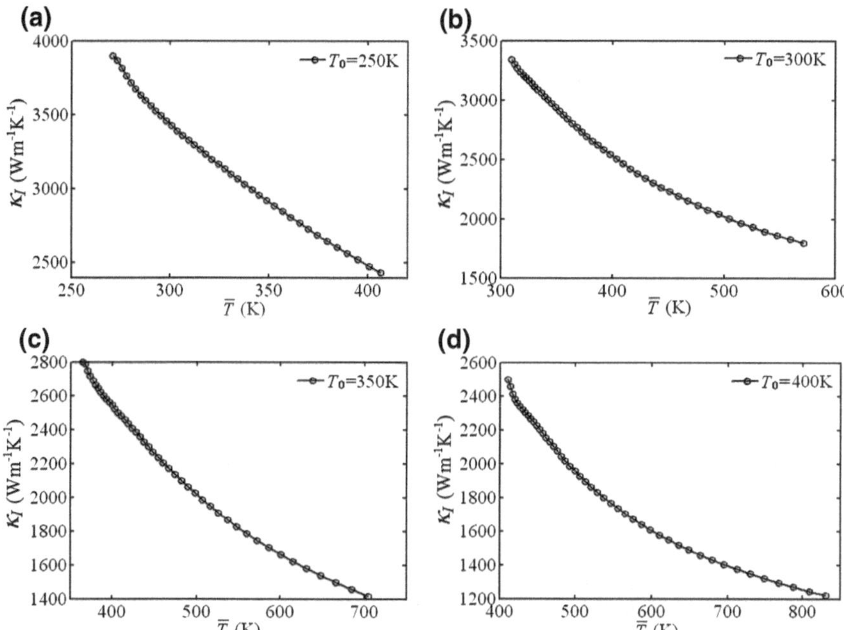

Fig. 2.28 Intrinsic thermal conductivities of SWCNT at different ambient temperatures. **a** Intrinsic thermal conductivity of SWCNT at 250K. **b** Intrinsic thermal conductivity of SWCNT at 300K. **c** Intrinsic thermal conductivity of SWCNT at 350K. **d** Intrinsic thermal conductivity of SWCNT at 400K. Reprinted from "Heat flow choking in carbon nanotubes, 53, Hai-Dong Wang, Bing-Yang Cao, Zeng-Yuan Guo, 1796–1800". Copyright [2010], with permission from Elsevier

and similarly, the second order coefficient is:

$$\beta_2 = \beta_2^* \frac{2\gamma\rho^3 C^4 T_0^2 L}{\kappa_I^2 c^2} \quad and \quad \beta_2^* = -\beta_{h2}^* \frac{T^2}{T_0^2} \tag{2.81}$$

As a result, a steady general heat conduction equation with a second TM resistance term is obtained as:

$$\kappa_I \left(1 - \frac{q^2}{2\gamma\rho^2 C^3 T^3}\right) \frac{d^2 T}{dx^2} + \left(1 + \frac{\varepsilon\kappa_I S}{\gamma\rho^2 C^3 T^2} - \beta_{h2}^* \frac{L^2 S}{T\kappa_I}\right) S \approx 0 \tag{2.82}$$

If $\beta_h^{**} = -\frac{8\varepsilon(1-\eta)}{\beta_S^{*2}} + \beta_{h2}^*$, where $\eta = \frac{\kappa_A}{\kappa_I} = 1 - \frac{q^2}{2\gamma\rho^2 C^3 T^3}$, $(0 \le \eta \le 1)$ is the ratio between the apparent and intrinsic thermal conductivities, the Eq. (2.82) can be written as:

$$\kappa_I \eta \frac{d^2 T}{dx^2} + \left(1 - \beta_h^{**} \beta_s^*\right) S \approx 0 \tag{2.83}$$

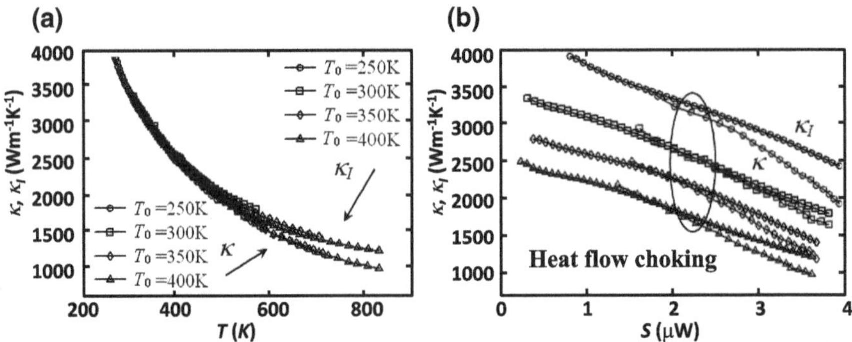

Fig. 2.29 Comparison between κ_I and κ_A of SWCNT. **a** Apparent and intrinsic thermal conductivities varies with temperature. **b** Apparent and intrinsic thermal conductivities varies with heating power. Reprinted from "Heat flow choking in carbon nanotubes, 53, Hai-Dong Wang, Bing-Yang Cao, Zeng-Yuan Guo, 1796–1800". Copyright [2010], with permission from Elsevier

when $\eta = 1$ and $\beta^*_{h2} = 0$, the Eq. (2.83) will reduce to the heat diffusion equation. Using the modified general heat conduction Eq. (2.83), the intrinsic thermal conductivity κ_I can be extracted from the experimental data from the Ref. [26] as shown in the Fig. 2.28.

The Fig. 2.28 shows that κ_I decreases as the temperature increases from 250 to 800 K. The apparent thermal conductivity κ_A can be extracted from Fourier's law, κ_I and κ_A are compared in the Fig. 2.29.

The circle in the Fig. 2.29b is drawn to mark the critical electrical power for heat flow choking. Below the critical electrical power, κ_I and κ_A are almost the same; above the critical power, κ_I is notably higher than κ_A and the difference is caused by the TM inertia effect. The comparison result highlights the potential application of the general heat conduction law in thermal analysis of the nanomaterials under the ultra-high heat flux conditions.

2.5 Conclusions

1. The relativistic mass of thermal energy of a single particle (molecule, atom, electron, etc.) is defined as thermon. All the thermons in a system form thermon gas, the heat flux can be understood as the direct flow of thermon gas.
2. Thermon gas is a kind of compressible fluid, whose state equations in ideal gas, dielectrics, and metals are given in this chapter. Similar to the state equation of ideal gas, a unified state equation exists for dielectrics and metals at the high temperature limit. At low temperatures, the difference in several state equations comes from the microscopic distribution functions.
3. Substituting the state equation of thermon gas into the momentum conservation equation, one can get a general heat conduction equation (law). It is a damped

wave equation that can be used to predict the propagation of thermal waves. The thermal wave is caused by the non-negligible temporal TM inertia.

4. A TSTM has been developed for metals heated by ultra-short pulsed lasers. This model is capable of predicting the transient unequilibrium heat conduction between electrons and lattices.
5. Under the ultra-high heat flux conditions, a steady non-Fourier heat conduction behavior is predicted by the TM theory for the first time, i.e., heat flow choking. Similar to the motion of air flow in a convergent nuzzle, when thermal Mach number equals unity, the heat flow choking occurs at the cold side of the heat conductor, causing a temperature jump. In this case, the intrinsic thermal conductivity is notably higher than the apparent one. The theoretical prediction is supported by the experimental result from literature.

References

1. J. Fourier, *Analytical Theory of Heat* (Dover Publications, New York, 1955)
2. G.V. Chester, A. Thellung, The law of Wiedemann and Franz. Proc. Phys. Soc. **77**(5), 309 (1961)
3. Z.Y. Guo, Movement and transport of thermomass–thermomass and thermon gas. J. Eng. Thermophys. **27**(4), 631–634 (2006). (in Chinese)
4. Z.Y. Guo, B.Y. Cao, H.Y. Zhu, Q.G. Zhang, State equation of phonon gas and its movement conservation equation. Acta Phys. Sinica **56**(6), 3306–3312 (2007)
5. A.L. Lavoisier, *Chemical Basis of Theory*. (Peking University Press, Beijing, 2008). (in Chinese)
6. I. Müller, *A History of Thermodynamics: The Doctrine of Energy and Entropy*. (Springer, New York, 2007)
7. W.F. Magie, *A Source Book in Physics* (McGraw-Hill, New York, 1935)
8. H.D. Wang, Z.Y. Guo, Thermon gas–heat carriers in gas and metals. Chin. Sci. Bull. **55**(1), 1–7 (2010)
9. L. Onsager, Reciprocal relations in irreversible processes. Phys. Rev. **37**, 405–426 (1931)
10. Q.G. Zhang, B.Y. Cao, Z.Y. Guo, Movement and transport of thermomass-state equation of thermon gas. J. Eng. Thermophys. **27**, 908–910 (2006). (in Chinese)
11. S.K. Ratkje, P.C. Hemmer, H. Holden, *The Collected Works of Lars Onsager: With Commentary*. (World Scientific, Singapore, 1996)
12. B.Y. Cao, Z.Y. Guo, Equation of motion of a phonon gas and non-Fourier heat conduction. J. Appl. Phys. **102**, 053503 (2007)
13. B.X. Cai, *The Basis of the Solid State Physics*. (Higher Education Press, Beijing, 1990). (in Chinese)
14. K. Huang, *Solid State Physics*. (Higher Education Press, Beijing, 1988). (in Chinese)
15. P. Mazur, Low-temperature specific heat of a thin film. Phys. Rev. B **23**(12), 6503–6511 (1981)
16. Y. Lu, Q.L. Song, S.H. Xia, Calculation of specific heat for aluminum thin films. Chin. Phys. Lett. **22**(9), 2346–2348 (2005)
17. W.H. Tang, R.Q. Zhang, *Equation of State Theory and Calculation Studies* (National University of Defense Technology Press, Beijing, 1999). (in Chinese)
18. M. Chester, Second sound in solids. Phys. Rev. **131**(5), 2013–2015 (1963)
19. W.Z. Dai, T.C. Niu, A finite difference scheme for solving a nonlinear hyperbolic two-step model in a double-layered thin film exposed to ultrashort-pulsed lasers with nonlinear interfacial conditions. Nonlinear Anal.: Hybrid Syst. **2**, 121–143 (2008)

20. S.I. Anisimov, B.L. Kapeliovich, T.L. Perelman, Electron emission from surface of metals induced by ultrashort laser pulses. Sov. Phys. JETP **39**, 375–377 (1974)
21. S. Lepri, R. Livi, A. Politi, Heat conduction in chains of nonlinear oscillators. Phys. Rev. Lett. **78**, 1896–1899 (1997)
22. O. Narayan, S. Ramaswarmy, Anomalous heat conduction in one dimensional momentum conserving systems. Phys. Rev. Lett. **89**, 200601 (2002)
23. S. Maruyama, A molecular dynamics simulation of heat conduction in finite length swnts. Phys. B: Condens. Matter **323**, 193–195 (2002)
24. S. Lepri, R. Livi, Heat in one dimension. Nature **421**, 327 (2003)
25. S. Sieniutycz, Relativistic thermohydrodynamics and conservation laws in continua with thermal inertia. Rep. Math. Phys. **49**(2–3), 361–370 (2002)
26. E. Pop, D. Mann, Q. Wang, K. Goodson, H. Dai, Thermal conductance of an individual single wall carbon nanotube above room temperature. Nano Lett. **6**(1), 96–100 (2006)
27. E. Pop, D. Mann, J. Reifenberg, K. Goodson, H.J. Dai, *Electro-Thermal Transport in Metallic Single-wall Carbon Nanotubes for Interconnect Applications* (International Electron Devices Meeting, Washington, DC, 2005)
28. E. Pop, D. Mann, J. Cao, Q. Wang, K. Goodson, H.J. Dai, Negative differential conductance and hot phonons in suspended nanotube molecular wires. Phys. Rev. Lett. **95**, 1555051 (2005)

Chapter 3
Experimental Investigation of Thermal Wave and Temperature Wave

Abstract The non-negligible thermomass inertia effect results in the non-Fourier heat conduction under the extreme conditions. In this chapter, the experimental study on the transient thermal wave and temperature wave will be discussed in detail. The propagation speed of temperature wave has been measured using a femtosecond laser thermoreflectance system. Meanwhile, the other parameters of the metallic nanofilms, such as electron–phonon coupling factor, thermal contact resistance, have been obtained.

3.1 Principles of Femtosecond Laser Thermoreflectance System

The femtosecond laser technique has been widely used in precise machining and manufacturing, high-density information storage, controlled thermonuclear fusion reactions, etc. Since the pulse duration of laser is at femtosecond level or even shorter, this technique has provided us a powerful tool to investigate the ultra-fast physical and chemical processes, known as the femtosecond physics and femtosecond chemistry [1]. Here the femtosecond laser transient thermoreflectance (TTR) system will be discussed.

3.1.1 Experimental Principle

Figure 3.1 shows the schematic diagram of the TTR system:

In a TTR system, a femtosecond laser is split into two beams: pump beam and probe beam, and the laser intensity of the probe beam is about 1/60 of that of the pump beam. The pump beam is used to heat the film sample, while the probe beam is used to detect the thermoreflectance signal. A half-wave plate is used to make the polarization direction of the probe beam perpendicular to that of the pump beam. By adjusting the optical axis direction of the Glan prism, the pump beam can be eliminated, letting the

H.-D. Wang, *Theoretical and Experimental Studies on Non-Fourier Heat Conduction Based on Thermomass Theory*, Springer Theses, DOI: 10.1007/978-3-642-53977-0_3,

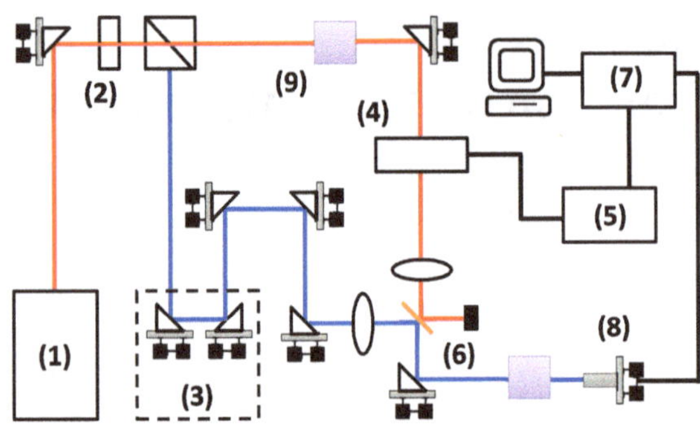

Fig. 3.1 Femtosecond laser transient thermoreflectance (TTR) system (rear heating-front detecting). Reprinted from "Theoretical and experimental study on the heat transport in metallic nanofilms heated by ultra-short pulsed laser, 54, Hai-Dong Wang, Wei-Gang Ma, Xing Zhang, Wei Wang, Zeng-Yuan Guo, 967–974". Copyright [2011], with permission from Elsevier

Table 3.1 Equipment specifications

Parameter	Value
Wavelength	800 (nm)
Pulse duration	80 (fs)
Pulse repetition frequency	80 (MHz)
Pump beam power	300 (mW)
Probe beam power	<5 (mW)
Diameter of focused laser spot	100 (μm)
Step distance	8.5 (nm)
Maximum stage distance	10 (cm)
Modulating frequency	1.2 (MHz)
Time constant	100 (ms)
Voltage detection range	1 (mV)

probe beam be detected only. A precise electrical stepping stage is used to control the optical paths of these two beams. In the experiment, the moving direction of the stage should be perfectly parallel with the optical path 1 μm step is corresponding to 6.67 fs time delay. A CCD camera is used to observe the focused spots of two beams, to make sure that the spots coincide with each other. The focused spot is about 100 μm measured in the optical image. An acoustic-optical modulator (AOM) is used to modulate the pump beam; the reference frequency is fed into a lock-in amplifier to select the thermoreflectance signal. The movement of stepping stage is controlled by the computer; a complete thermoreflectance versus time curve can be drawn by changing the optical path delay step by step. The specifications of the experimental equipment are listed in the Table 3.1:

Figure 3.2 shows the measurement principle of TTR system. When the film sample is heated by pump beam pulses, the surface temperature will change periodically. The time delay between two neighboring pump pulses is 13 ns, which is far larger than the pulse duration 80 fs. As a result, the temperature rise caused by one pulse will decrease to zero before the next coming pulse. The surface reflectance is related

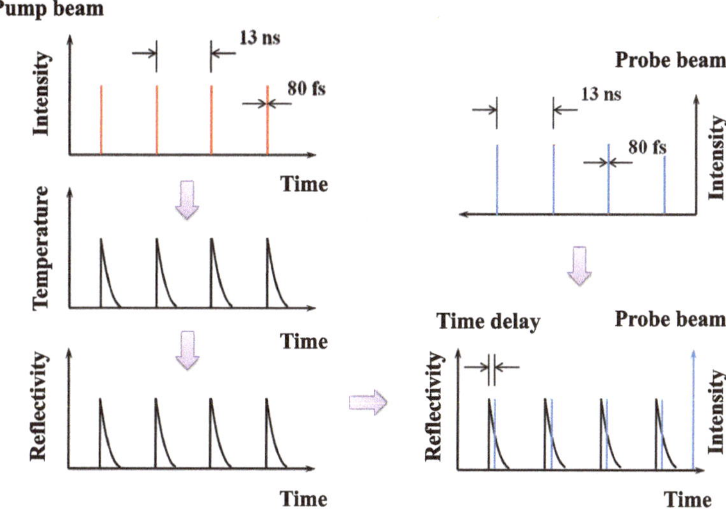

Fig. 3.2 Measurement principle of TTR system

to the dielectric function $\varepsilon(\omega)$, which can be expressed as [2]:

$$\varepsilon(\omega) = \varepsilon_1(\omega) + i\varepsilon_2(\omega), \tag{3.1}$$

where ε_1 and ε_2 are the real and imaginary parts of the function. The dielectric function is decided not only by the incident photon frequency, but also by the electron and lattice temperatures. The change of dielectric function causes the change of reflectance [3]. For the photons with shorter wavelength, the interband transition of electrons plays a dominant role for the change of reflectance; for the photons with longer wavelength in our case, the intraband transition is also important [4]. The linear relationship between the reflectance and temperature is given as [2]:

$$\Delta R = a\Delta T_e + b\Delta T_l, \tag{3.2}$$

where a and b are the constants for electron temperature and lattice temperature. Because the electron temperature (~120 K) is usually much higher than the lattice temperature (~1 K), the Eq. 3.2 can be simplified as $\Delta R = a\Delta T_e$.

In the Fig. 3.2, the change of temperature results in the change of surface reflectance. The time delay between the pump and probe beams is precisely controlled by the stepping stage. In order to monitor the weak thermoreflectance signal, the lock-in amplifier technique is used in the experiment. AOM is used to produce a modulated laser pulse train, where a high-frequency ultrasonic wave is fed into a crystal and causes time-dependent variation of refractive index. It is much like a plane grating for incident laser beam, results in Raman-Nuys diffraction [5]. The first-order diffraction light is used as the pump beam, whose intensity is $I_1 = I\sin^2(\mu/2)$, where I is the intensity of incident light and μ is the Raman-Nuys coefficient. The diffraction efficiency could be more than 95 % for a well-designed AOM. Changing the power of ultrasonic wave, the diffraction efficiency and intensity of output light are also changed.

In a lock-in amplifier, the phase sensitive detection (PSD) is to select the useful signals according to the reference frequency [6]. For example, the measured signal is $V_E \sin(\omega_R t + \theta_E)$ and the reference signal is $V_R \sin(\omega_R t + \theta_R)$, where V_E and V_R are the signal amplitudes, ω_R is the reference frequency, θ_E and θ_R are the phases. After the processing of a multiplier, the output signal is:

$$\begin{aligned} V_1 &= V_E V_R \sin(\omega_R t + \theta_E) \sin(\omega_R t + \theta_R) \\ &= \tfrac{1}{2} V_E V_R \cos(\theta_R - \theta_E) + \tfrac{1}{2} V_E V_R \sin(2\omega_R t + \theta_E + \theta_R) \end{aligned} \tag{3.3}$$

where $1/2 V_E V_R \cos(\theta_R - \theta_E)$ is the direct current (DC) component. The $2\omega_R$ component can be eliminated using a high-frequency filter. After the processing of a mixer using $V_R \sin(\omega_R t + \theta_R \pi/2)$ as reference signal, the output signal is

$$V_2 = \frac{1}{2} V_E V_R \cos\left(\theta_R - \theta_E - \frac{\pi}{2}\right) + \frac{1}{2} V_E V_R \sin\left(2\omega_R t + \theta_E + \theta_R - \frac{\pi}{2}\right) \tag{3.4}$$

After high-frequency filtering, one can get:

$$V_2 = \frac{1}{2} V_E V_R \cos\left(\theta_R - \theta_E - \frac{\pi}{2}\right) = \frac{1}{2} V_E V_R \sin(\theta_R - \theta_E) \tag{3.5}$$

Combining the results of multiplier and mixer, the measured amplitude and phase are:

$$V_E = \frac{2}{V_R} \sqrt{V_1^2 + V_2^2} \tag{3.6}$$

$$\theta_R - \theta_E = \tan^{-1}\left(\frac{V_2}{V_1}\right) \tag{3.7}$$

The X, Y components are:

$$X_E = V_E \cos(\theta_R - \theta_E) \tag{3.8}$$

$$Y_E = V_E \sin(\theta_R - \theta_E) \tag{3.9}$$

3.1.2 Experimental Setup

The main parts of femtosecond laser TTR system are listed below:

Figure 3.3 shows the optical system in the experiment. There are four types of mirrors used: the light splitter is used to split the light into the pump and probe

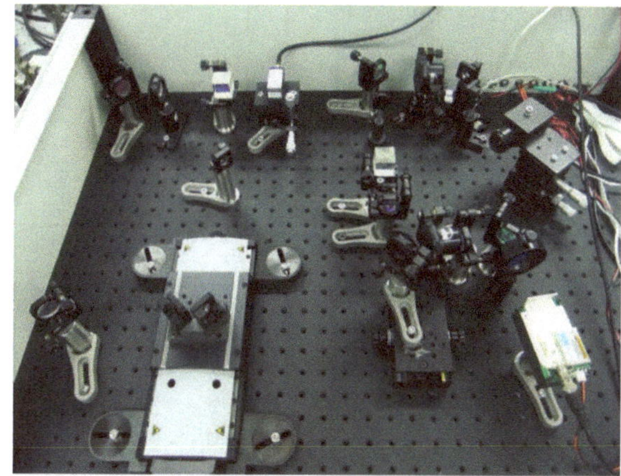

Fig. 3.3 Optical system of the femtosecond laser TTR experiment

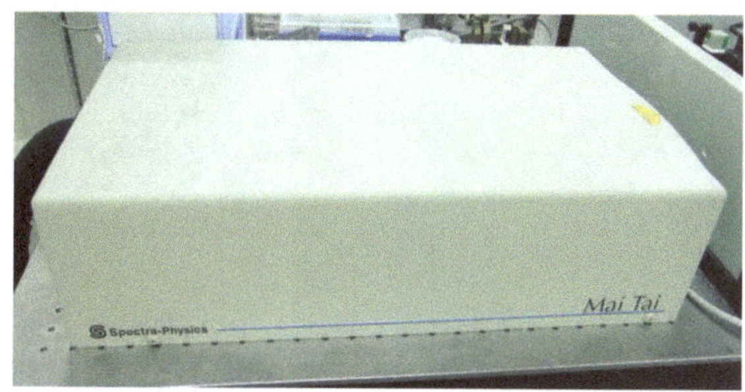

Fig. 3.4 Femtosecond laser

beams; the totally reflecting prisms are used to reflect the beams; the Glan prisms are used to eliminate the pump beam, leaving only the probe beam to pass through; the biconvex lens are used to focus the laser beam on the sample surface. Every prism has two rotational degrees of freedom and can be moved in x, y, and z directions.

Figure 3.4 shows the Titanium–Sapphire femtosecond laser from Spectra-Physics Inc. The controlled wavelength range is from 720 nm to 930 nm. A singlehanded water cooling system is used to cool the laser. In order to maintain a constant environmental temperature and humidity, a clean room has been built to place the whole TTR system.

Figure 3.5 shows the Stanford Research System SR844 lock-in amplifier used in the experiment. The detecting signal range is from 25 kHz to 200 MHz.

Figure 3.6 shows the PI M-410 precise stepping stage. A single minimum moving step is 8.5 nm.

Figure 3.7 shows the M080 acoustic-optical modulator (AOM) from Gooch & Housego Inc. The modulation efficiency is more than 85 %.

Figure 3.8 shows the C5331 electrophotonic detector from Hamamatsu Inc. It is used to convert the weak optical signal into the detectable voltage signal. The resolution is 4.5×10^4 VW^{-1}. Besides the equipments shown in the Figs. 3.3, 3.4,

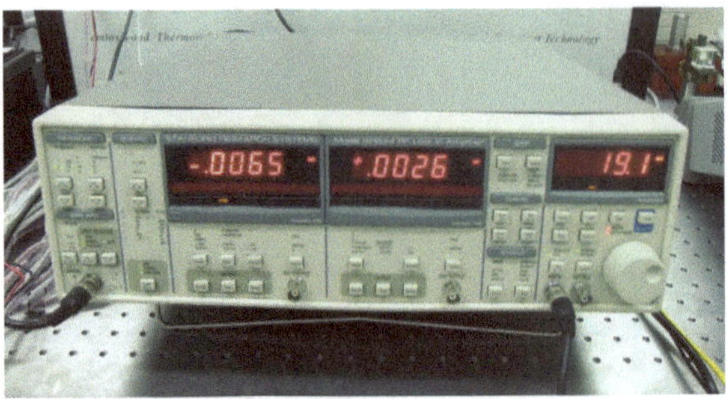

Fig. 3.5 Lock-in amplifier

Fig. 3.6 Precise stepping stage

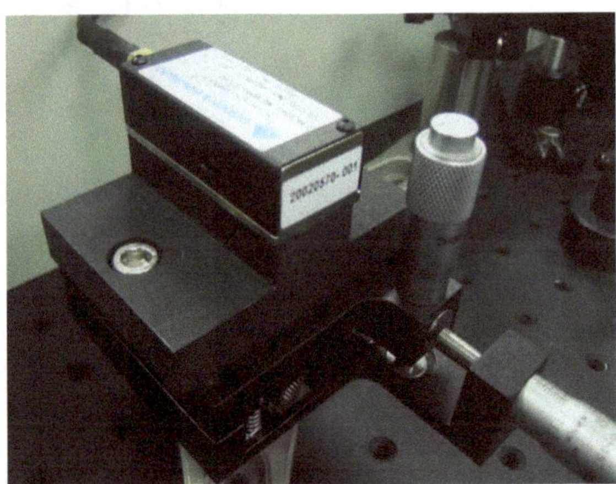

Fig. 3.7 Acoustic-optical modulator (AOM)

3.5, 3.6, 3.7 and 3.8, the signal generator, oscilloscope, DC power supply are also needed in the experiments.

In order to successfully detect the femtosecond TTR signals, some key questions should be paid special attention to:

1. Determination of the zero optical path position. The zero optical path position is the zero point of TTR signal varied with respect to time. The TTR signal can be only detected around the zero point. At the zero position, the optical paths of

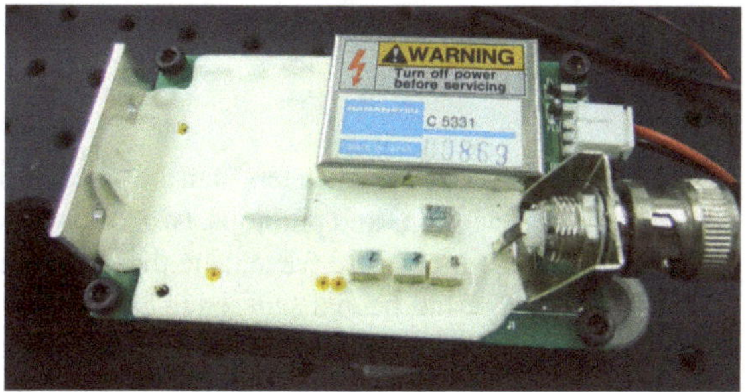

Fig. 3.8 Electrophotonic detector

the pump and probe beams are exactly the same. In the experiment, a nonlinear optical beta-BaB_2O_4 crystal (BBO crystal) is used to find the zero position. Both the pump and probe beams are focused on the BBO crystal, if there is no time delay between these two beams, the coherent conditions are satisfied and a third doubled-frequency laser spot will appear between the pump and probe laser spots. In this case, the pump and probe pulses arrive at the BBO crystal surface at the same time, the corresponding position on the stepping stage is referred to as the zero position.

2. Adjustment of the parallelism of the stepping stage. In order to make sure that the focused pump and probe laser spots coincide with each other on the surface of the film sample, the moving direction of the stepping stage has to be parallel with the light. In the experiment, the reflected light from the prisms on the stepping stage is shooting at a screen which is $4 \sim 5$ far away. Even a slight deviation of the parallelism will cause a significant movement of the laser spot on the screen. By carefully adjusting the angles of the reflecting prisms on the stage, the laser spot on the screen stays at the same place while the stage is moving.
3. Elimination of the electrical and optical noises. In order to increase the signal-to-noise ratio, the electrical and optical noises should be eliminated in the experiment. The electrical noise mainly comes from the signal generator and AOM; the power supply and connection cable of the electrophotonic detector should be separated from these devices. The optical noise mainly comes from the scattered light of the mirrors and prisms. The electrophotonic detector is placed in a black metal box to avoid the effect of scattered light. Meanwhile, an optical filter is placed in front of the detector to decrease the leaked pump light. The intensity ratio between the probe and pump beams is controlled to be below 1/30.

3.2 Thermal Wave and Temperature Wave in Metallic Nanofilms

The femtosecond laser TTR system can be used to investigate the ultra-fast heat transfer between electrons and phonons, two kinds of wave phenomena are involved: thermal wave and temperature wave. The theoretical analysis is given first.

The thermal wave has been discussed in Sect. 2.2.4, because of the non-negligible temporal thermomass inertia effect, heat is dissipated in the form of waves. It is much like the propagation of mechanical waves or water waves, the thermal wave can be reflected at the solid boundaries. As a matter of fact, there is another temperature wave existed, it is induced by the periodic temperature boundary conditions. Eckert gives the governing equation for the heat conduction in semi-infinite solid material as [7]:

$$\frac{\partial T}{\partial t} = \alpha \frac{\partial^2 T}{\partial x^2}, \tag{3.10}$$

$$\text{when } t = 0, \quad T = 0; \quad \text{at } x = 0, \quad T_0 = f(t). \tag{3.11}$$

where α is the thermal diffusivity of the material. Assuming that the boundary temperature is a cosine function, the analytical solution of the Eq. 3.10 is

$$T = \frac{A_0}{2} + T_{0M} \exp\left(-\sqrt{\frac{\pi}{\alpha\tau_0}}x\right)\cos\left(\frac{2\pi t}{\tau_0} - \sqrt{\frac{\pi}{\alpha\tau_0}}x\right) \tag{3.12}$$

where $A/2$ and T_{0M} are the average temperature of the surface and the maximum amplitude of temperature oscillation. The Fig. 3.9 compares the temperature waves at the front surface and at the x distance. It is seen that a time delay of $\Delta t = 0.5x(\tau/\alpha\pi)^{1/2}$ exists between these two waves, where τ is the period of temperature wave. The wavelength of temperature wave x is given as

$$x_0 = 2\sqrt{\pi\alpha\tau_0} \tag{3.13}$$

Based on the above analysis, the propagation speed of temperature wave can be obtained as:

$$v = \frac{\Delta x}{\Delta t} = 2\sqrt{\frac{\pi\alpha}{\tau_0}} \tag{3.14}$$

The propagation speed is decided by the thermal diffusivity and the period of boundary temperature.

It should be noted that the thermal wave and temperature wave both appear as the periodic temperature oscillation, but the physical mechanisms are quite different. The temperature wave is caused by the periodic temperature change at the boundaries, the heat conduction is still governed by diffusion equations. The thermal wave is governed by hyperbolic heat conduction equations, it is a transient non-Fourier heat

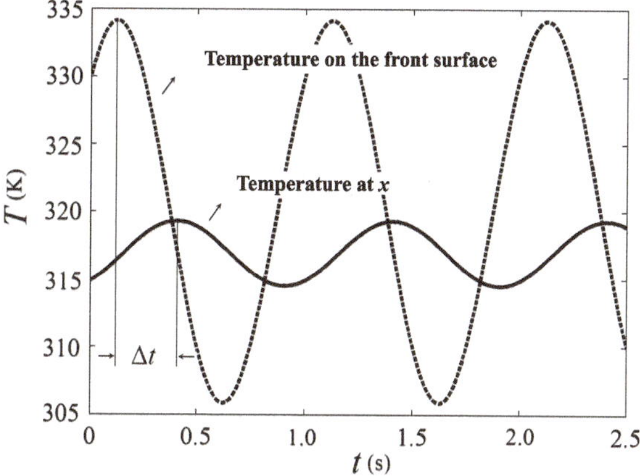

Fig. 3.9 Periodic temperature wave. Reprinted from "Theoretical and experimental study on the heat transport in metallic nanofilms heated by ultra-short pulsed laser, 54, Hai-Dong Wang, Wei-Gang Ma, Xing Zhang, Wei Wang, Zeng-Yuan Guo, 967–974". Copyright [2011], with permission from Elsevier

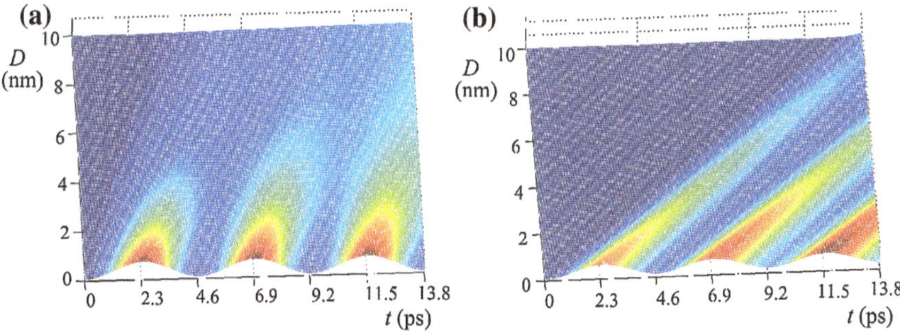

Fig. 3.10 Comparison between the temperature wave **a** and thermal wave **b**

conduction phenomenon. The physical essence of thermal wave is the propagation of thermomass pressure wave in solids. The propagation speed of thermal wave is only decided by the material properties, especially the characteristic times [8–10] To summarize, the thermal wave is "wave" while the temperature wave is "diffusion."

Figure 3.10 gives the comparison between the propagation of temperature wave and thermal wave in materials where the horizontal ordinate is time and the vertical ordinate is propagation distance, the different colors represent different electron temperatures. It is seen that the temperature field of thermal wave has obvious "characteristic lines," whose slope is the propagation speed of thermal wave. On the other hand, the temperature field of temperature wave has obvious features of heat diffusion. A one-dimensional damped hyperbolic equation is given as:

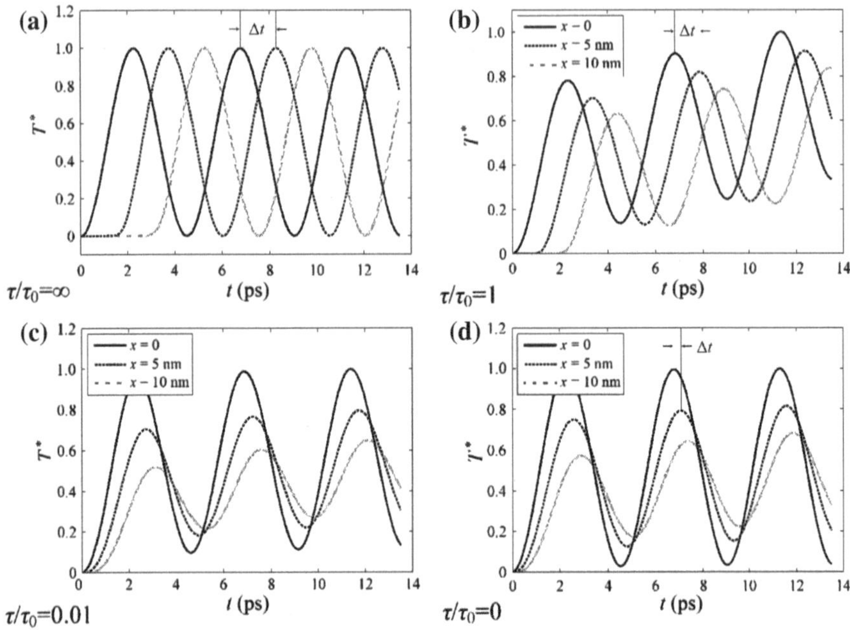

Fig. 3.11 Temperature versus time curves plotted with respect to the different characteristic times. Reprinted from "Theoretical and experimental study on the heat transport in metallic nanofilms heated by ultra-short pulsed laser, 54, Hai-Dong Wang, Wei-Gang Ma, Xing Zhang, Wei Wang, Zeng-Yuan Guo, 967–974". Copyright [2011], with permission from Elsevier

$$\tau\frac{\partial^2 T}{\partial t^2}+\frac{\partial T}{\partial t}=\frac{\kappa}{\rho C}\frac{\partial^2 T}{\partial x^2} \tag{3.15}$$

where the relative importance of fluctuation term versus diffusion term depends on the ratio between τ and τ_0. When $\tau = \infty$, the Eq. 3.15 reduces to an undamped hyperbolic equation; when $\tau = 0$, the Eq. 3.15 reduces to a pure heat diffusion equation.

Figure 3.11 shows the transformation from thermal wave to temperature wave, where $\tau/\tau_0 = \infty$, 1, 0.01, and 0 from the Fig. 3.11 (a) to the Fig. 3.11 (d), T^* is the normalized temperature. It can be concluded that:

1. When $\tau/\tau_0 = \infty$, the Eq. 3.15 becomes an undamped one. Then as the τ/τ_0 decreases, the diffusion term becomes more and more important, the wave amplitude decreases as well.
2. When $\tau/\tau_0 = 0$, the Eq. 3.15 becomes pure diffusive and the thermal wave disappears. In this case, the temperature oscillation is referred to as the temperature wave. There is still a time delay existed for the temperature peaks at different propagation distances. The propagation speed of temperature wave can be calculated as $\Delta x/\Delta t$, where Δx is the distance and Δt is the corresponding time delay.

There are three typical transient heat conduction behaviors that exist for different characteristic times as shown in the Fig. 3.12:

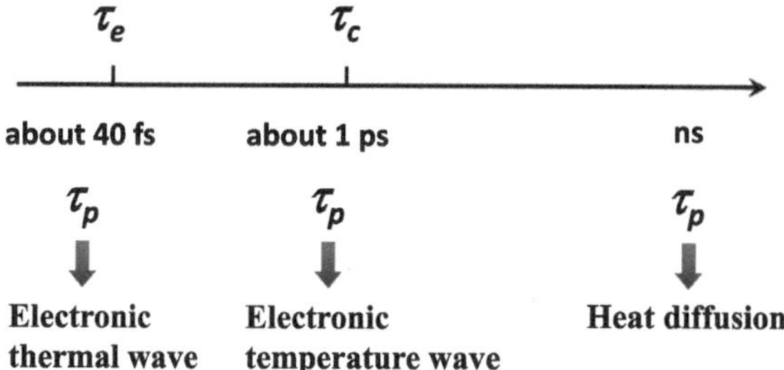

Fig. 3.12 Three typical transient heat conduction behaviors. Reprinted from "Theoretical and experimental study on the heat transport in metallic nanofilms heated by ultra-short pulsed laser, 54, Hai-Dong Wang, Wei-Gang Ma, Xing Zhang, Wei Wang, Zeng-Yuan Guo, 967–974". Copyright [2011], with permission from Elsevier

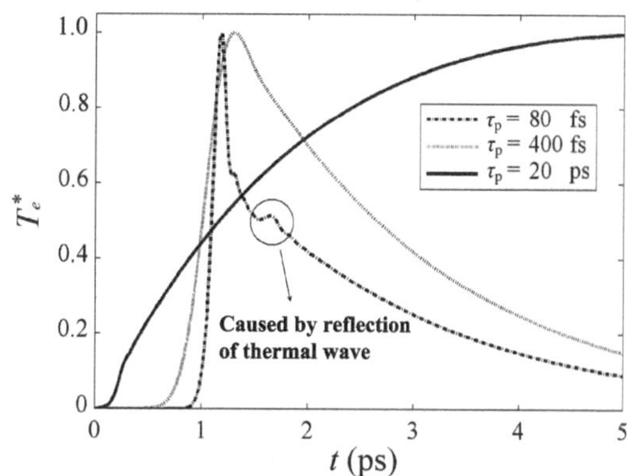

Fig. 3.13 Transient heat conduction at different τ_p. Reprinted from "Theoretical and experimental study on the heat transport in metallic nanofilms heated by ultra-short pulsed laser, 54, Hai-Dong Wang, Wei-Gang Ma, Xing Zhang, Wei Wang, Zeng-Yuan Guo, 967–974". Copyright [2011], with permission from Elsevier

In the Fig. 3.12, there are three typical characteristic times: electron characteristic time τ_e, lattice thermalization time τ_c, and laser pulse duration τ_p. As τ_p increases, the heat conduction in metals experiences three processes: thermal wave, temperature wave, and heat diffusion in thermal equilibrium state. The Fig. 3.13 gives the calculation examples with different τ_p:

As shown in the Fig. 3.13, when $\tau_e \sim \tau_p = 80$ fs, the thermal wave exists in metals and the reflected thermal wave from the rear surface will cause a second temperature peak at the front surface. In this case, the propagation speed of thermal wave is $v = (\kappa_e/C\tau_p)^{0.5} = 4.52 \times 10^5$ ms^{-1}. When $\tau_c \sim \tau_p = 400$ fs, the heat diffusion equation governs and the temperature wave occurs instead of the thermal wave. In this case, the propagation speed of temperature wave is $v = 2(\pi\kappa_e/C\tau_p)^{0.5} = 7.16 \times 10^5$ ms^{-1}, which is significantly larger than the speed of thermal wave, close to the Fermi speed When $\tau_c << \tau_p = 20$ ps the electrons are in thermal equilibrium state with lattices, the electron temperature equals the lattice temperature. Based on the Fig. 3.13, it is seen that the thermal wave can be reflected from the boundary interface, while the temperature wave cannot. The propagation speed of thermal wave is decided by the material properties only, while the speed of temperature wave is affected by the temperature boundary conditions.

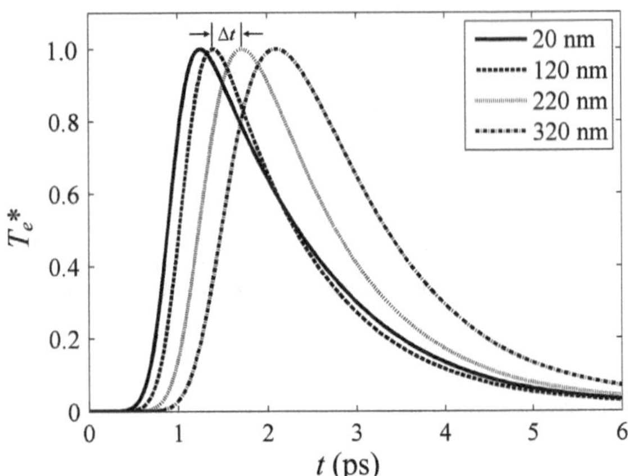

Fig. 3.14 Propagation of temperature wave in metallic films. Reprinted from "Theoretical and experimental study on the heat transport in metallic nanofilms heated by ultra-short pulsed laser, 54, Hai-Dong Wang, Wei-Gang Ma, Xing Zhang, Wei Wang, Zeng-Yuan Guo, 967–974". Copyright [2011], with permission from Elsevier

3.3 Measurement of Temperature Wave in Metallic Nanofilms

The femtosecond laser TTR system is used to measure the propagation speed of temperature wave in metal films. A rear heating-front detecting method is used in the experiment. For theoretical predictions, a second-order accurate finite difference scheme is used to solve the heat conduction equation.

In the Fig. 3.14, T_e* is the normalized electron temperature, Δt is the time delay due to different film thicknesses. As discussed above, the propagation speed of temperature wave is $v = \Delta x/\Delta t$.

Figure 3.15 shows the varied propagation speed plotted with respect to time (or diffusion depth $\Delta x = v\Delta t$). $t = 0$ is the time when the laser pulse arrives at the film surface. $\delta = 15.3$ nm is the penetration depth of laser in gold.

It is seen in the Fig. 3.15 that the propagation speed decreases as time increases, gradually reaches a constant value at 0.7 ps. There are two regions involved: (A) the electrons and lattices are in the unequilibrium thermal state, the heat transfer takes place in electrons while the lattice temperature almost remains a constant. In this case, the propagation speed is very fast, $v_e = 2 \times (\pi\alpha_e/\tau_p)^{1/2} = 7 \times 10^5$ ms^{-1} ($\tau_p \sim 400$ fs); (B) the electron–phonon coupling plays an important role, the energy of electrons is transferred to lattices by collisions. The coupling effect limits the propagation of temperature wave greatly, causing a decreasing speed. Finally, the speed reaches a constant, $v_e = 2 \times (\pi\alpha_e/\tau_c)^{1/2} = 2.5 \times 10^5$ ms^{-1} ($\tau_c \sim 3$ ps).

The propagation of temperature wave in the regions (A) and (B) are investigated separately. For the region (A), there are three gold (Au) film samples prepared: 27.2 nm, 39.9 nm, and 55.5 nm The measured result is shown in the Fig. 3.16:

The Fig. 3.16 shows the original measured thermoreflectance signals, T_e* is the normalized electron temperature. The measured data are smoothed using a wavelet filter, the final result is shown in the Fig. 3.17:

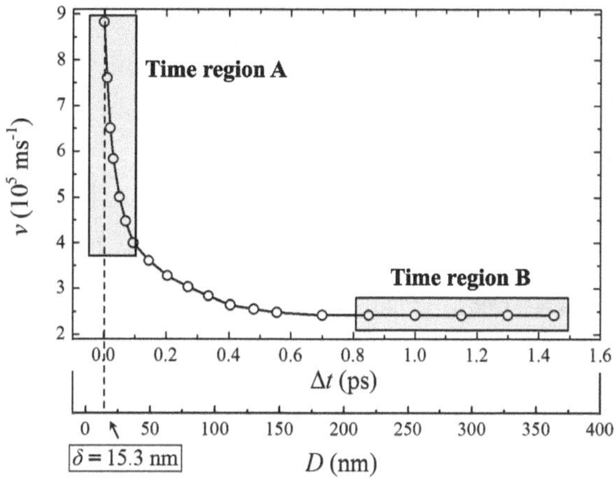

Fig. 3.15 Propagation speed of temperature wave plotted with respect to time (or diffusion depth). Reprinted from "Theoretical and experimental study on the heat transport in metallic nanofilms heated by ultra-short pulsed laser, 54, Hai-Dong Wang, Wei-Gang Ma, Xing Zhang, Wei Wang, Zeng-Yuan Guo, 967–974". Copyright [2011], with permission from Elsevier

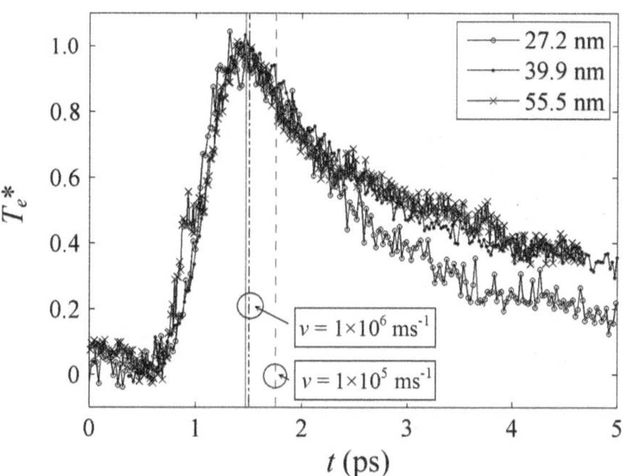

Fig. 3.16 Thermoreflectance signals of different Au film samples. Reprinted from "Theoretical and experimental study on the heat transport in metallic nanofilms heated by ultra-short pulsed laser, 54, Hai-Dong Wang, Wei-Gang Ma, Xing Zhang, Wei Wang, Zeng-Yuan Guo, 967–974". Copyright [2011], with permission from Elsevier

Figure 3.17 shows the smoothed electron temperature curve, in the zoom-in inset, it is seen that the temperature peak of thicker film is delayed in time. The linear fitting result in the Fig. 3.18 gives the propagation speed.

In the Fig. 3.18, the slope is the propagation speed of temperature wave in Au films. The minimum step of moving stage is 3.4 μm and the corresponding time resolution is 23 fs, which is shown as the error bar in the Fig. 3.18. The linear fitted propagation speed is 5.2×10^5 ms^{-1} $\sim1.4\times10^6$ ms^{-1}, the best-fitted value is 8.1×10^5 ms^{-1}, which agrees with the theoretical prediction of region (A) in the Fig. 3.15. For very thin Au films, the electrons are the energy carriers, the propagation speed is close to the Fermi speed. In Fig. 3.16, two propagation speeds of 1×10^5 ms^{-1}and 1×10^6 ms^{-1} are drawn for comparison, it demonstrates that the measurement uncertainty may cause some deviation to the exact value of v, but its order of magnitude is known for certain.

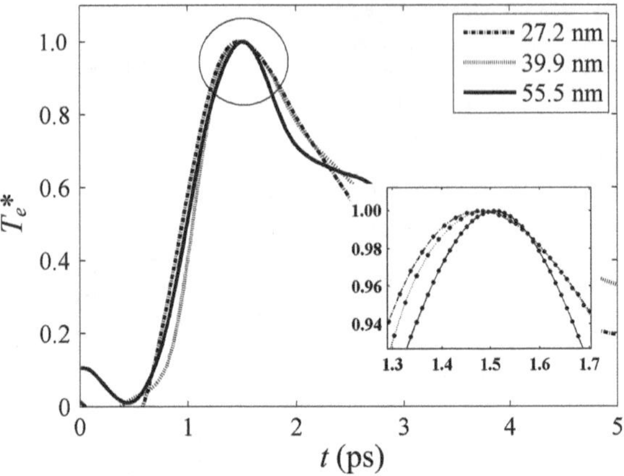

Fig. 3.17 Smoothed electron temperature curve plotted with respect to time. Reprinted from "Theoretical and experimental study on the heat transport in metallic nanofilms heated by ultra-short pulsed laser, 54, Hai-Dong Wang, Wei-Gang Ma, Xing Zhang, Wei Wang, Zeng-Yuan Guo, 967–974". Copyright [2011], with permission from Elsevier

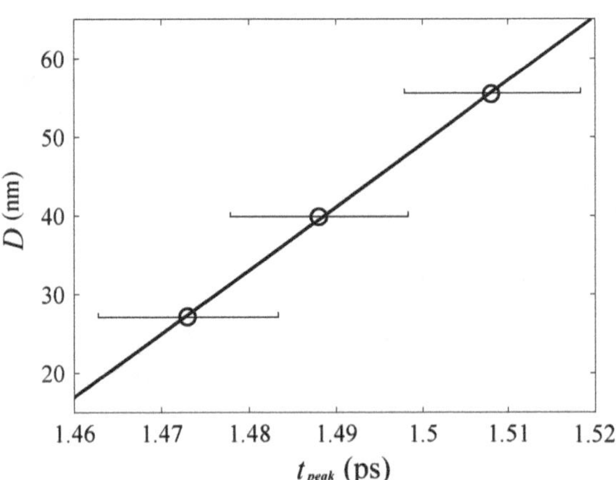

Fig. 3.18 Linear fitting result of film thickness versus temperature peak time. Reprinted from "Theoretical and experimental study on the heat transport in metallic nanofilms heated by ultra-short pulsed laser, 54, Hai-Dong Wang, Wei-Gang Ma, Xing Zhang, Wei Wang, Zeng-Yuan Guo, 967–974". Copyright [2011], with permission from Elsevier

The experimental results from Ref. [11] can be used to evaluate the propagation speed in the region (B) of the Fig. 3.15. The same experimental method is used in this work and in Ref. [11], the main difference is that thicker Au film samples (100 nm, 200 nm, and 300 nm) are measured in Ref. [11].

Figure 3.19 is the original picture in Ref. [11]. It is seen that the thick film has an obvious time delay in temperature peak due to the propagation of temperature wave, the speed is about $3.1 \times 10^5 \mathrm{ms}^{-1}$. The theoretical prediction for region (B) in the Fig. 3.15 is $v_e = 2 \times (\pi \alpha_e/\tau_c)^{1/2} = 2.5 \times 10^5\ \mathrm{ms}^{-1}$, which agrees with the experimental result. It has been concluded that the two regions for temperature propagation in metals are proved by the experiments, the electron–phonon interactions will decrease the propagation speed.

In this thesis, the observed temperature oscillation has been proved to be the temperature wave, not the thermal wave. Only when the pulse duration is close to the characteristic time of electrons, the thermal wave may be observed. Our work clears

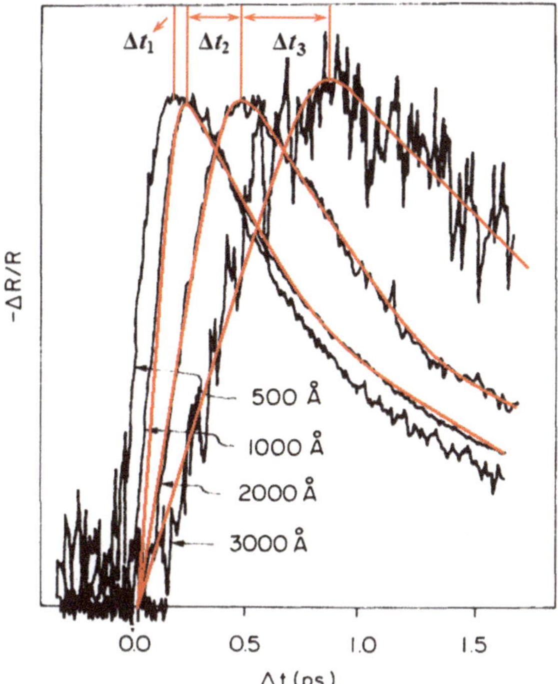

Fig. 3.19 Experimental results of Au film samples with different thicknesses [11]. Reprinted from "Theoretical and experimental study on the heat transport in metallic nanofilms heated by ultra-short pulsed laser, 54, Hai-Dong Wang, Wei-Gang Ma, Xing Zhang, Wei Wang, Zeng-Yuan Guo, 967–974". Copyright [2011], with permission from Elsevier

up a misunderstanding that the temperature oscillation does not have to be the thermal wave, it could be the temperature wave caused by heat diffusion in unequilibrium thermal state.

3.4 Electron–Phonon Coupling Factor and Interfacial Thermal Resistance

An important parameter in the two-temperature model is electron–phonon coupling factor G, which represents the strength of interactions between the electrons and phonons. Here, we measured the factor G of Au nanofilms on different substrates. Meanwhile, the thermal contact resistance R and thermal conductivity of substrate κ_s can be obtained. Using a genetic algorithm (GA), several parameters can be decided simultaneously.

Some researchers have studied the femtosecond laser processing mechanisms in glass and alloys [12–14], but the accurate measurement of thermal properties during the transient laser heating is still lacking. The coupling factor G and thermal contact resistance R are coupled in the experiment, usually R has to be neglected in order to get G. Here, a new method has been developed to determine these two parameters

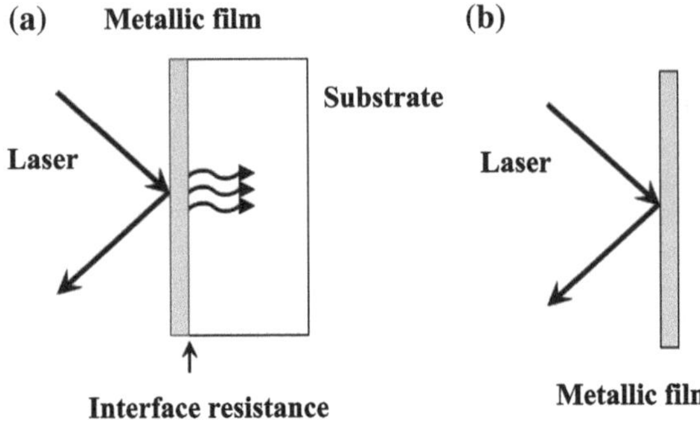

Fig. 3.20 Comparison between the two-layer model and single-layer model

at femtosecond level. A two-layer model is used for theoretical analysis as shown in the Fig. 3.20:

Figures 3.20a, b are the two-layer model and single-layer model, respectively. For the two-layer model, G and R are unknown and coupled parameter; for the single-layer model, R is ignored and G becomes the only unknown parameter.

$$-K_e \frac{\partial T_e}{\partial x}\bigg|_{x=d} = -K_s \frac{\partial T_s}{\partial x}\bigg|_{x=d} \tag{3.16}$$

$$-K_e \frac{\partial T_e}{\partial x}\bigg|_{x=d} = \frac{T_e - T_s}{R}\bigg|_{x=d} \tag{3.17}$$

where $x = d$ is the interface between the film and substrate, κ_s, T_s, and R are the thermal conductivity and temperature of substrate, thermal contact resistance, respectively.

In 1986, Balageas gives an analytical solution to the two-layer model as [15]:

$$V_D(t) = 1 + 2\frac{\sum_{i=1}^{2} x_i w_i}{\sum_{i=1}^{2} x_i} \sum_{K=1}^{\infty} \frac{\sum_{i=1}^{2} x_i \cos(w_i \gamma_K)}{\sum_{i=1}^{2} x_i w_i \cos(w_i \gamma_K)} \exp\left(-\frac{\gamma_K^2 t}{\eta_2^2}\right) \tag{3.18}$$

where $e_i = \sqrt{\kappa_i \rho_i C_i}$, $\eta_i = \frac{L_i}{\sqrt{\alpha_i}}$, $(i = 1, 2)$, κ ρ, C, and L are the thermal conductivity, density, specific heat, and thickness, respectively. i is the number of layer. $x_i = \frac{e_1}{e_2} - (-1)^i$, $w_i = \frac{\eta_1}{\eta_2} - (-1)^i$. γ_K is the number K real root of the equation $\sum_{i=1}^{2} x_i \sin(w_i \gamma) = 0$. For non-Dirac laser distribution, $V(t) = \int_0^t \varphi(t')V_D(t - t')dt'$, where the laser pulse shape function $\phi\ (t)$ is considered.

In 1994, Hui and Tan [16] gave a transmission line model suitable for n layer films. In this model, the laser was taken as a constant-current source, the thermal resistance was taken as the electrical resistance and the temperature was taken as

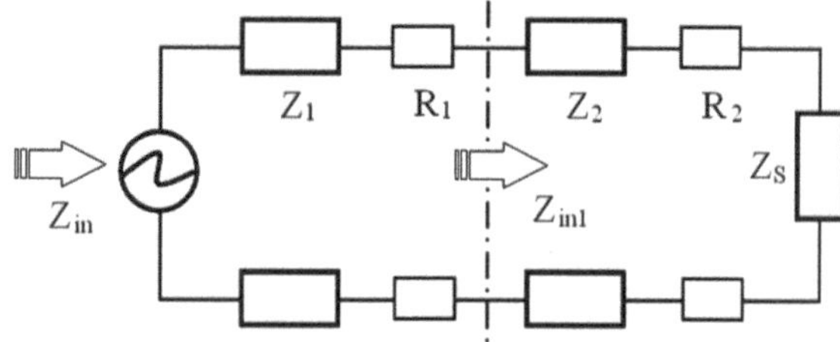

Fig. 3.21 Transmission line model

the voltage. A two-layer example of the transmission line model is given in the Fig. 3.21:

In the Fig. 3.21, Z and R are the film impedance and contact impedance, respectively. Z_{in} is the total impedance on the right hand side. According to the transmission line model, one can get

$$Z_{in1} = Z_2 \frac{(Z_S + R_2) + Z_2 \tanh(\gamma_2 L_2)}{Z_2 + (Z_S + R_2) \tanh(\gamma_2 L_2)}, \tag{3.19}$$

$$Z_{in} = Z_1 \frac{(Z_{in1} + R_1) + Z_1 \tanh(\gamma_1 L_1)}{Z_1 + (Z_{in1} + R_1) \tanh(\gamma_1 L_1)}, \tag{3.20}$$

$$\varepsilon(s) = Z_{in} Q(s), \tag{3.21}$$

$$Q(s) = \frac{\sqrt{\pi}}{2a}(1 - R)J \exp\left(\frac{\left(s - 2a^2 t_0\right)^2}{4a^2} - a^2 t_0^2\right) erfc\left(\frac{s}{2a} - at_0\right), \tag{3.22}$$

$$a^2 = \frac{4 \ln 2}{t_p^2}, \tag{3.23}$$

where s is the Laplace operator, $\varepsilon\ (s)$ and $Q(s)$ are the Laplace expressions for the front surface temperature and laser heat flux. t_0, t_p, J, and R are the laser pulse time delay, pulse duration, power density, and reflectance, respectively. The impedance is $Z = \frac{1}{\kappa\gamma}$, where $\gamma = \sqrt{\frac{s\rho C}{\kappa}}$. Finally the front surface temperature can be calculated through the anti-Laplace transform of $\varepsilon\ (s)$ as:

$$T(t) = \frac{1}{2\pi i} \int_{\sigma - i\infty}^{\sigma + i\infty} \varepsilon(s) \exp(st) ds, \ s = \sigma + iw, \ t > 0. \tag{3.24}$$

The transmission line model can be widely used for a multilayer film, but the anti-Laplace transform is difficult to carry out.

When $G = 0$ in the parabolic twostep (PTS) model, it reduces to a pure heat diffusion model. The numerical calculation result is compared with the analytical solutions as shown in the Fig. 3.22.

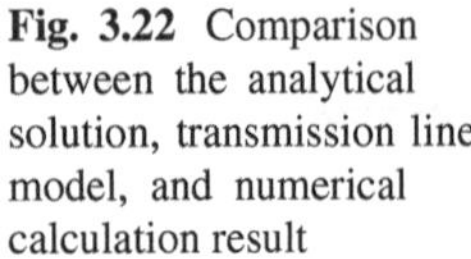

Fig. 3.22 Comparison between the analytical solution, transmission line model, and numerical calculation result

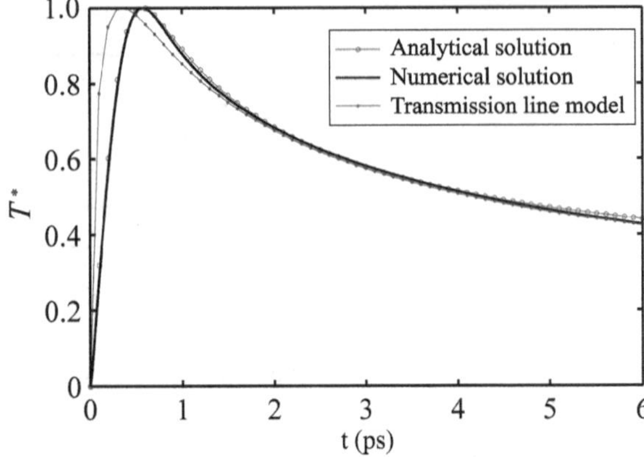

Fig. 3.23 Normalized electron temperatures plotted with respect to time at different contact resistances

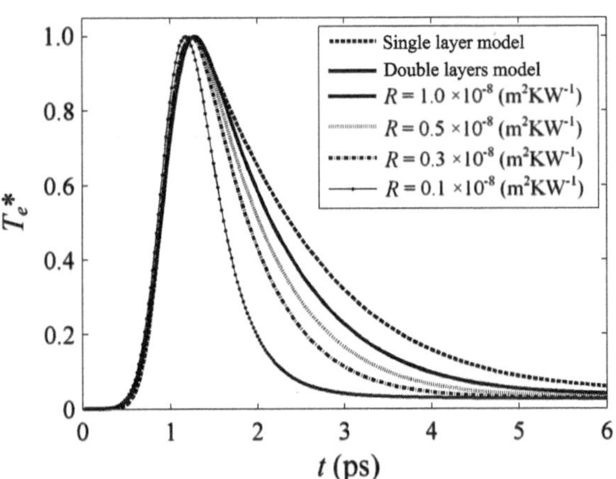

It is seen in Fig. 3.22 that the numerical result agrees well with the analytical result and the decreasing curve of the transmission line model. The deviation of the transmission line model is induced by the anti-Laplace transform The numerical solution has been proved to be accurate for two-layer model, it will be used for the theoretical predictions.

Figure 3.23 gives the normalized electron temperature calculated at different contact resistances. T_e* decreases more rapidly when the contact resistance decreases. Here R is the thermal contact resistance for transient electron energy exchange, differing from the resistance in steady state.

Chen [17] developed a temperature-dependent model for electron–phonon coupling, pointing out that the factor G increases as the electron temperature increases. A decreasing R will decrease the electron temperature, thus the factor G decreases as well. In this way, G and R are connected and should be solved together. In the single-layer model, R is ignored and G will be overestimated, this is the so-called apparent coupling factor in Ref. [3]. Hopkins and Norris [4] developed a three-temperature

Fig. 3.24 AFM images of Au nanofilms with different thicknesses

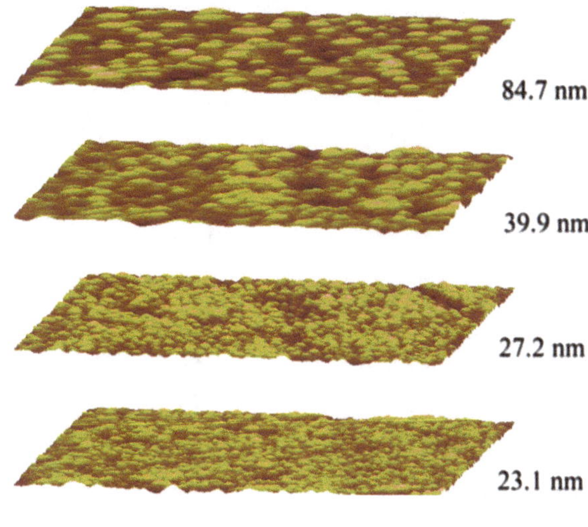

Table 3.2 Measured average grain sizes of different Au nanofilms

Thickness (nm)	23.1	27.2	39.9	84.7
Average grain size (nm)	15	23	30	49

model, considering the electron temperature, lattice temperature, and substrate temperature at the same time. In this model, the decreasing electron temperature is caused by both the electron–phonon coupling and energy exchange at the interface. Some other researchers pointed out that the electrons transferred the energy to the phonons inside the metallic films first, then this part of energy was transferred to the substrate through the phonon–phonon interactions [18–20].

In order to obtain G and R simultaneously, GA method is used. The GA method is a multiparameter optimization method, which is widely applied to solve the complicated nonlinear and multidimensional problems. The other gradient-based methods are difficult to carried out for these complicated optimization problems [21]. The GA method does not depend on the derivative of the objective parameters or the initial guess values [22]. The GA method is developed based on the genetic transmission and natural selection rules [23]. The objective function is $S = \sum_{i=1}^{N} [\tilde{\theta}_i(t) - \theta_i(t)]^2$, where $\tilde{\theta}(t)$ and $\theta(t)$ are the measured and calculated normalized temperatures. The minimum objective function is the minimum variance between $\tilde{\theta}(t)$ and $\theta(t)$. A classical process of GA method includes four operations: selection, crossover, mutation, and elitism. After many generations, the best selected parameters give the minimum value of the objective function [14].

In the experiment, the Au films are prepared using a physical vapor deposition method. The substrates are 500 μm borosilicate glass and 100 nm SiC layer. The fabricated Au nanofilms have been scanned using an atomic force microscopy (AFM), the results are shown in the Fig. 3.24:

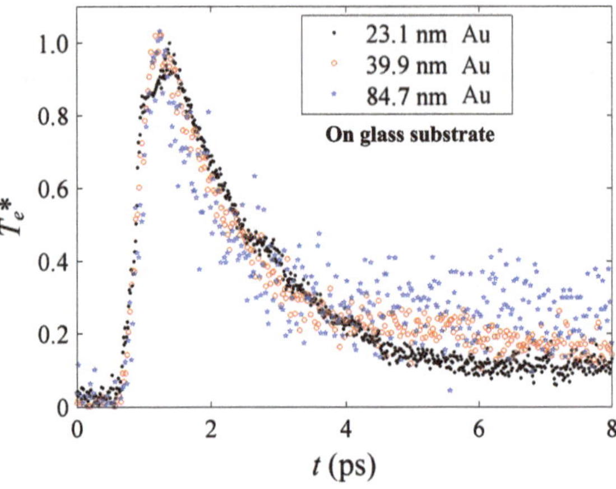

Fig. 3.25 Normalized electron temperatures of Au films with different thicknesses

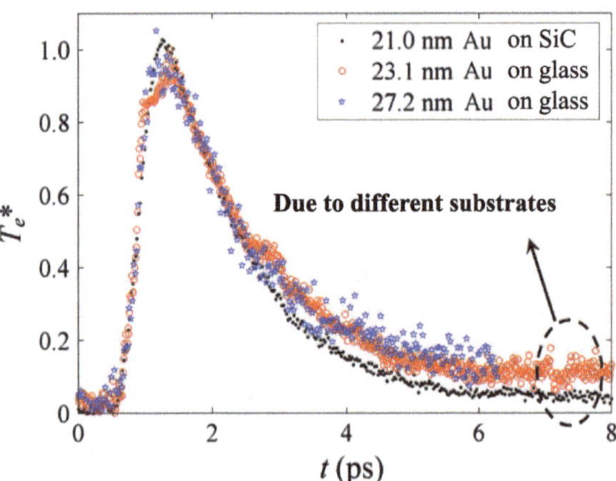

Fig. 3.26 Normalized electron temperatures of Au films deposited on different substrates

The measured average grain size of Au nanofilms are listed in the Table 3.2:

It is noted that the average grain size increases as the thickness increases. The deposition time of thick film is larger than that of thin film, making the grains to grow larger. This can be clearly seen in the Fig. 3.24. But for the thick film, the surface roughness is also large, thus the thermoreflectance signal could be decreased.

Figure 3.25 is the measured normalized electron temperature curves of different Au film samples. The measured data coincide with each other in the rising edge, while a deviation can be observed in the decreasing curve after 4 ps. For thick films, the electron temperature rise is relatively smaller, and the difference between the temperatures at peak and at smooth section is smaller as well. To draw the normalized temperature curves according to the peak value $T_e* = T_e/T_{emax}$, the smooth section of thick film is relatively higher. On the other hand, the experimental data of thick film are more scattered, because of its larger surface roughness.

Table 3.3 Calculation results of the GA method

No.	1	2	3	4	5	Average
$G \times 10^{-16}$ ($Wm^{-3}K^{-1}$)	2.47	2.47	2.57	2.56	2.57	2.53
$R \times 10^{8}$ (m^2KW^{-1})	0.80	0.81	1.1	1.1	0.78	0.92

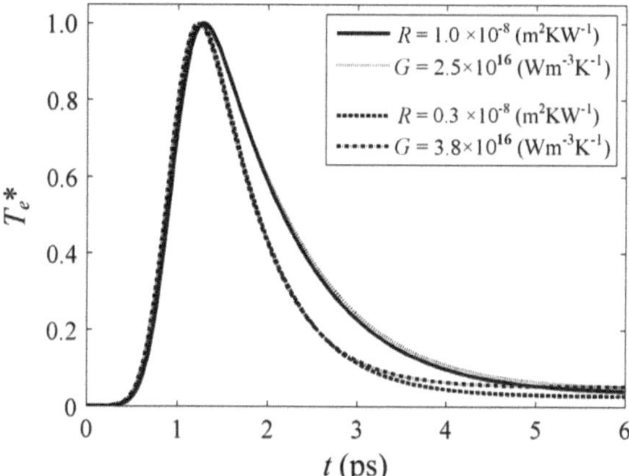

Fig. 3.27 Sensitivity analysis of electron temperature to G and R

Figure 3.26 is the measured normalized electron temperature curves of Au films on different substrates. The film thicknesses are almost the same. It is seen in the Fig. 3.26 that the temperature curve of 21 nm Au film on SiC substrate is relatively lower than that on glass substrate. It implies that the energy of electrons is more readily transferred into the substrate with high thermal conductivity. By analyzing the decreasing electron temperature curve, the thermal conductivity of the substrate can be obtained.

Before using the GA method to analyze the experimental data, its validation should be proved at first. Here we give the correct values of G and R: $G = 2.5\times10^{16}$ $Wm^{-3}K^{-1}$, $R = 1.0\times10^{-8}$ m^2KW^{-1}. Substituting G and R into the PTS model, one can get a normalized electron temperature curve, which can be seen as the "experimental data." Using the GA method to fit the experimental data, G and R can be determined. If the GA method is accurate, the recalculated G and R should be the same as the given correct values. The Table 3.3 gives the calculation results of five rounds:

Based on the calculation results in Table 3.3, the GA method is applicable with the relative uncertainties of 1.2 % for G and 8 % for R. R is fitted with higher uncertainty due to its lower sensitivity to the electron temperature. For example, assuming $G = 2.5\times10^{16}\,Wm^{-3}K^{-1}, R = 1.0\times10^{-8}\,m^2\,KW^{-1}$ for curve (A) and $G = 3.8\times10^{16}\,Wm^{-3}K^{-1}$, $R = 0.3\times10^{-8}m^2\,KW^{-1}$ for curve (B), the calculated electron temperature curves are shown in the Fig. 3.27:

In the Fig. 3.27, it is noted the electron temperature curve calculated with 52 % increased G coincides with that calculated with 70 % decreased R. This means that

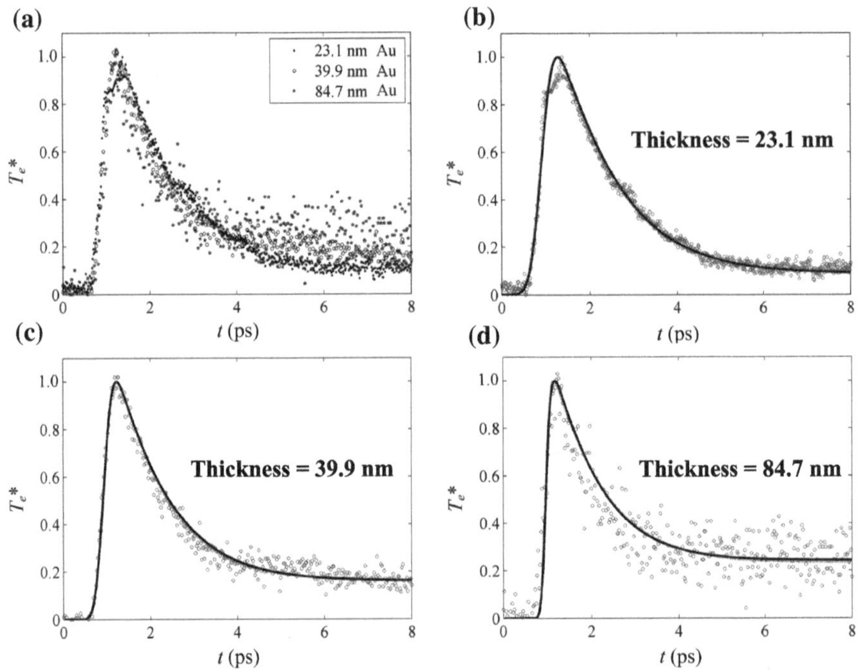

Fig. 3.28 Best-fitting electron temperature curves for Au films on glass substrate

the decreasing electron temperature curve is more sensitive to G, thus the uncertainty of G is relatively smaller.

Figure 3.28 gives the best-fitting electron temperature curves for Au films. Besides the parameter G and R, the other material properties are chosen to be $\kappa_e = \kappa T_e / T_l$ (κ = 315 Wm^{-1}K^{-1}), $C_e = A_e T_e$ (A_e= 70 Jm^{-3}K^{-2}), $C_l = 2.5\times10^6$ Jm^{-3}K^{-1}. The specific heat and thermal conductivity of glass substrate are $C_s = 1.73\times10^6$ Jm^{-3}K^{-1}, κ_s = 1.10 Wm^{-1}K^{-1}. The best-fitted values of G and R using the GA method are listed in the Table 3.4:

The results in Table 3.4 are calculated based on the two-layer model. Using the single-layer model, the coupling factor G are: 1.95×10^{16} Wm^{-3}K^{-1}(23.1 nm film), 2.10×10^{16} Wm^{-3}K^{-1}(39.9 nm film), and 2.20×10^{16} Wm^{-3}K^{-1}(84.7 nm film). It is noted that G of single-layer model is larger than G of two-layer model. Because the single-layer model ignores the energy exchange with the substrate, the phonon coupling becomes the only way to cool the electrons.

Figure 3.29 shows the measured coupling factor G plotted with respect to the film thickness. It is seen that the data of single-layer model are larger than the data of two-layer model. On the other hand, the measured data are close to the bulk value, showing no significant thickness dependence. The bulk value in the Fig. 3.29 is calculated using the Eq. 3.25 proposed by Qiu and Tien [24]:

Table 3.4 Best-fitted results of Au films with different thicknesses

No.	Parameter	23.1 nm	39.9 nm	84.7 nm
1	$G \times 10^{-16}$ (Wm^{-3}K^{-1})	1.7394	1.9727	2.0869
	$R \times 10^{8}$ (m^{2}KW^{-1})	5.91	12.88	29.80
2	$G \times 10^{-16}$ (Wm^{-3}K^{-1})	1.7394	1.9020	1.8606
	$R \times 10^{8}$ (m^{2}KW^{-1})	5.64	18.82	21.82
3	$G \times 10^{-16}$ (Wm^{-3}K^{-1})	1.7333	2.0434	1.9657
	$R \times 10^{8}$ (m^{2}KW^{-1})	5.73	17.83	24.85
4	$G \times 10^{-16}$ (Wm^{-3}K^{-1})	1.7273	2.1000	1.9899
	$R \times 10^{8}$ (m^{2}KW^{-1})	5.36	16.84	27.58
5	$G \times 10^{-16}$ (Wm^{-3}K^{-1})	1.7455	1.9727	2.2000
	$R \times 10^{8}$ (m^{2}KW^{-1})	5.82	12.88	23.94
Average	$G \times 10^{-16}$ (Wm^{-3}K^{-1})	1.74	2.00	2.02
	$R \times 10^{8}$ (m^{2}KW^{-1})	5.69	15.85	25.60

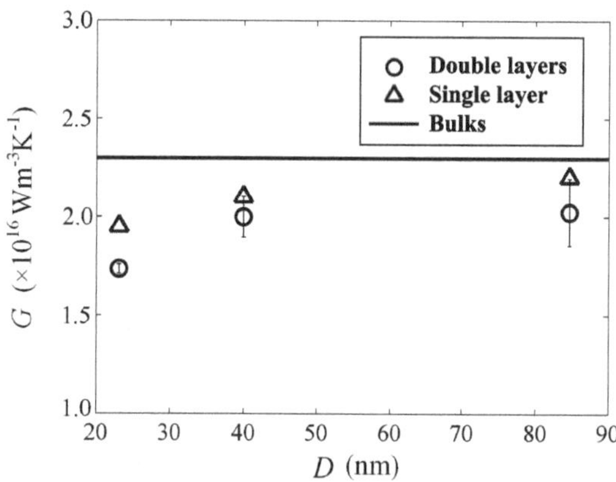

Fig. 3.29 Electron–phonon coupling factor plotted with respect to the film thickness

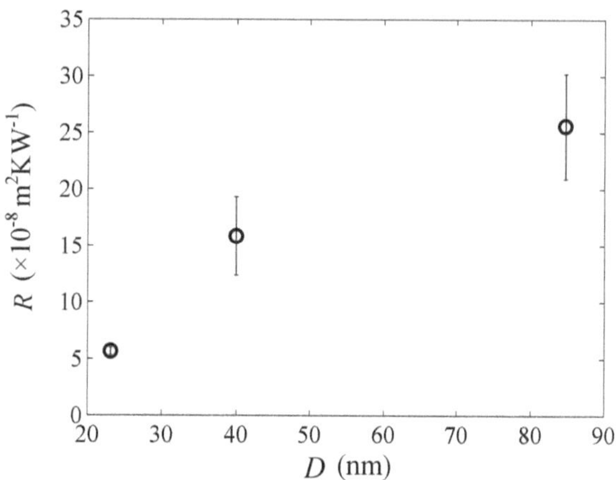

Fig. 3.30 Thermal contact resistance plotted with respect to the film thickness

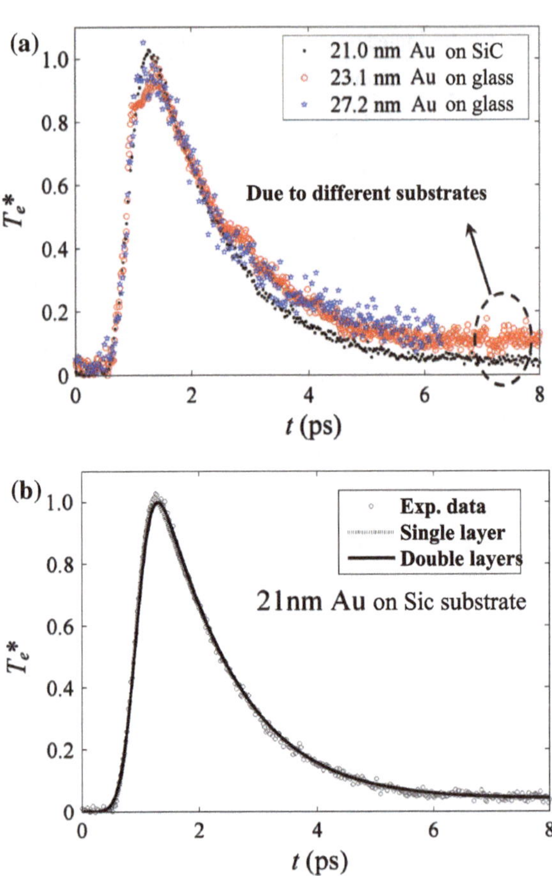

Fig. 3.31 Best-fitted curve for the electron temperature of Au film on SiC substrate

$$G = \frac{9}{16}\frac{nk_B^2 T_D^2 v_F}{\Lambda T_l E_F} \tag{3.25}$$

where Λ is the mean free path of electrons, n, k_B, T_D, v_F, and E_F are the electron number density, Boltzmann constant, Debye temperature, Fermi velocity, and Fermi energy, respectively. The predicted value of Eq. 3.25 is 2.3×10^{16} $\text{Wm}^{-3}\text{K}^{-1}$, slightly larger than the experimental data.

Figure 3.30 shows the measured thermal contact resistance plotted with respect to the film thickness. Majumdar and Reddy [18] proposed a physical model to calculate the thermal resistances of metallic films for transient laser heating:

$$R = \frac{1}{\sqrt{Gk_l}} + R_{ps} = R_{ep} + R_{ps}; \quad R_{ep} = \frac{1}{\sqrt{Gk_l}} \tag{3.26}$$

where R_{ep} and R_{ps} are the thermal resistances induced by electron–phonon and phonon–substrate interactions. The total thermal resistance R is the sum of these two resistances. According to the Eq. 3.26, R_{ep} equals 0.3×1^{-8} m^2KW^{-1}, which is only 10 % of the total resistance. In other words, R_{ps} plays an important role.

Our experimental result (5×10^{-8} m^2KW^{-1} $\sim26\times10^{-8}$ m^2KW^{-1}) is close to the thermal resistance on SiO_2 substrate in Ref. [14], since the thermal conductivities of SiO_2 and glass are very close. It is seen in the Fig. 3.30 that the thermal contact resistance R increases as the film thickness increases R depends on the interface roughness and thermal stress of the metallic film. The roughness increases as the film thickness increases, and the thermal stress is proportional to the deposition time of film. As a result, R is directly related with the film thickness [25, 26]. On the other hand, decreasing the surface roughness and thermal stress will decrease the thermal contact resistance.

Figure 3.31 shows the best-fitted curve for the electron temperature. The thickness of SiC layer is 100 nm, its thermal conductivity could be evaluated from the fitted temperature curve. Here, three parameters G, R, and κ_s are set as objective parameters for calculation. According to the best-fitted result using the GA method, one can get $G = 1.93\times10^{16}$ $Wm^{-3}K^{-1}$, $R = 11.14\times10^{-8}$ m^2KW^{-1}, and $\kappa_s = 131.30$ $Wm^{-1}K^{-1}$. The thermal conductivity of SiC layer is much larger than that of glass substrate, so the electron temperature decreases more rapidly on the SiC substrate. Meanwhile, the coupling factor G on the SiC substrate is slightly larger. It demonstrates that the substrate with high thermal conductivity is better for cooling the electron temperature.

3.5 Conclusions

1. A femtosecond laser TTR system has been established. We have overcome many key problems, such as: complicated light path adjustment, elimination of electrical and optical noises, selection of weak thermoreflectance signals, etc. The time resolution is at femtosecond level.
2. Two wave phenomena have been predicted for transient laser heating, i.e., thermal wave and temperature wave. The physical essence of thermal wave is the thermomass pressure wave in solids, while the temperature wave is caused by the periodic temperature condition based on the heat diffusion model. The propagation speed of temperature wave is decided by the material properties and boundary conditions.
3. The governing equation for transient heat conduction in metals is a damped hyperbolic equation. The relative importance of wave term or diffusion term is decided by three characteristic times, i.e., electron relaxation time τ_e, lattice thermalization time τ_c, and laser pulse duration τ_p.
4. The propagation speed of temperature wave is measured using a rear heating-front detecting method. The pulse duration is about 400 fs. The measured propagation speed is close to the Fermi speed, agreeing with the theoretical prediction. The experimental result in Ref. [11] is the propagation of temperature wave in electron–phonon coupling region, not the thermal wave.

5. Using the GA method, the coupling factor G, thermal contact resistance R, and substrate thermal conductivity κ_s can be determined simultaneously. The measured G is independent of the film thickness, while the measured R increases as the film thickness increases, due to the increasing interface roughness and thermal stress.

References

1. G. Khitrova, P.R. Berman, M. Sargent, Theory of pump-probe spectroscopy. J. Opt. Soc. Am. B **5**(1), 160–170 (1988)
2. P.E Hopkins, P.M. Norris , L.M. Phinney, S.A. Policastro, R.G. Kelly, Thermal conductivity in nanoporous gold films during electron-phonon nonequilibrium. J. Nanomater. **2008**, 418050 (2008)
3. R. Rosei, D.W. Lynch, Thermomodulation spectra of al, au, and cu. Phys. Rev. B **5**(10), 3883–3894 (1972)
4. P.E. Hopkins, P.M. Norris, Substrate influence in electron-phonon coupling measurements in thin au films. Appl. Surf. Sci. **253**, 6289–6294 (2007)
5. K. Huang, *Solid State Physics* (Higher Education Press, Beijing, 1988) (in Chinese)
6. J.S. Zhang, J.S. Zheng, X.P. Wan, *Lock-in techniques* (Xian University of Electronic Science and Technology of China Press, Xian, 1994) (in Chinese)
7. E.R.G. Eckert, *Heat and Mass Transfer* (McGraw-Hill, New York, 1959)
8. H.Y. Zhang, S.G. Wu, Femtosecond laser induced film damage mechanical process. Acta Phys. Sin. **56**(9), 5314–5317 (2007)
9. M.J. Maurer, Relaxation model for heat conduction in metals. J. Appl. Phys. **40**(13), 5123–5130 (1969)
10. M.J. Maurer, H.A. Thompson, Non-fourier effects at high heat flux. J. Heat Transfer **95**(2), 284–286 (1973)
11. S.D. Brorson, J.G. Fujimoto, E.P. Ippen, Femtosecond electronic heat-transport dynamics in thin gold-films. Phys. Rev. Lett. **59**(17), 1962–1965 (1987)
12. S.M. Lee, D.G. Cahill, Heat transport in thin dielectric films. J. Appl. Phys. **81**(6), 2590–2595 (1997)
13. A.R. Joseph, C.L. John, D.J. Stephen, Subsurface damage in some single crystalline optical materials. Appl. Opt. **44**(12), 2241–2249 (2005)
14. S. Orain, Y. Scudeller, S. Garcia, T. Brousse, Use of genetic algorithms for the simultaneous estimation of thin films thermal conductivity and contact resistances. Int. J. Heat Mass Transfer **44**(20), 3973–3984 (1972)
15. D.L. Balageas, J.C. Krapez, P. Cielo, Pulsed photothermal modeling of layered materials. J. Appl. Phys. **59**(2), 348–357 (1986)
16. P. Hui, H.S. Tan, A transmission-line theory for heat conduction in multilayer thin films. Compon. Packag. Manuf. Technol. Part B: Adv. Packag. **17**(3), 426–434 (1994)
17. J.K. Chen, W.P. Latham, J.E. Beraun, The role of electroncphonon coupling in ultrafast laser heating. J. Laser Appl. **17**(1), 63–68 (2005)
18. A. Majumdar, P. Reddy, Role of electroncphonon coupling in thermal conductance of metalc-nonmetal interfaces. Appl. Phys. Lett. **84**(23), 4768–4770 (2004)
19. Z.B. Ge, D.G. Cahill, P.V. Braun, Thermal conductance of hydrophilic and hydrophobic interfaces. Phys. Rev. Lett. **96**(18), 186101 (2006)
20. M.L. Roukes, M.R. Freeman, R.S. Germain et al., Hot electrons and energy transport in metals at millikelvin temperatures. Phys. Rev. Lett. **55**(4), 422–425 (1985)
21. J.V. Beck, K.J. Arnold, *Parameter Estimation in Engineering and Science* (Wiley, New York, 1977)

22. S. Garcia, J. Guynn, E.P. Scott, Use of genetic algorithms in thermal property estimation: Part ii simultaneous estimation of thermal properties. Numer. Heat Transfer **33**, 149–168 (1998)
23. D.E. Goldberg, *Genetic Algorithms in Search, Optimization and Machine Learning* (Addison-Wesley, MA, 1989)
24. T.Q. Qiu, C.L. Tien, Heat transfer mechanisms during short-pulse laser heating of metals. J. Heat Transfer **115**, 835–841 (1993)
25. M.I. Flik, P.E. Phelan, C.L. Tien, Thermal model for the bolometric response of high tc superconducting films to optical pulses. Cryogenics **30**(12), 1118–1128 (1990)
26. D. Sakami, A. Lahmar, Y. Scudeller, F. Danes, J.P. Bardon, Thermal contact resistance and adhesion studies on thin copper films on alumina substrates. J. Adhes. Sci. Technol. **15**(12), 1403–1416 (2001)

Chapter 4
Experimental Proof of Steady-State Non-Fourier Heat Conduction

Abstract It was discussed in Chap. 2 that the non-negligible thermomass inertia will cause non-Fourier heat conduction in steady states, for example, the heat flow choking phenomenon. In this chapter, the experimental evidence is given for the steady non-Fourier heat conduction under the ultra-high heat flux and low temperature conditions. As the foundation of the theoretical prediction, the electrical and thermal conductivities of the metallic nanofilms have been accurately measured in a wide temperature range. Meanwhile, the breakdown of Wiedemann–Franz law at low temperatures is observed in the experiment.

4.1 Electrical and Thermal Conductivities of Metallic Nanofilms

The metallic nanofilms are used to observe the non-Fourier heat conduction since the heat flux could exceed 10^{10} Wm^{-2} at low temperatures. Because the size effect plays an important role for nanofilms, especially at low temperatures, the electrical and thermal conductivities should be accurately determined by the experiment.

4.1.1 Direct Current Heating Experiment of Metallic Nanofilms

Figure 4.1 shows the typical fabrication procedures of the nanofilm samples [1]: (1) The Si substrate is oxidized and covered with a layer of 180 nm thick SiO_2 and then 320 nm thick electron beam (EB) resist is spin-coated on the top; (2) The pattern of nanofilm is directly drawn on the EB resist using the EB lithography; (3) Using a physical vapor deposition (PVD) method, the metallic film is deposited on the substrate with a 5 nm thick adhesion layer of titanium; (4) The chip is immersed in the

H.-D. Wang, *Theoretical and Experimental Studies on Non-Fourier Heat Conduction Based on Thermomass Theory*, Springer Theses, DOI: 10.1007/978-3-642-53977-0_4,

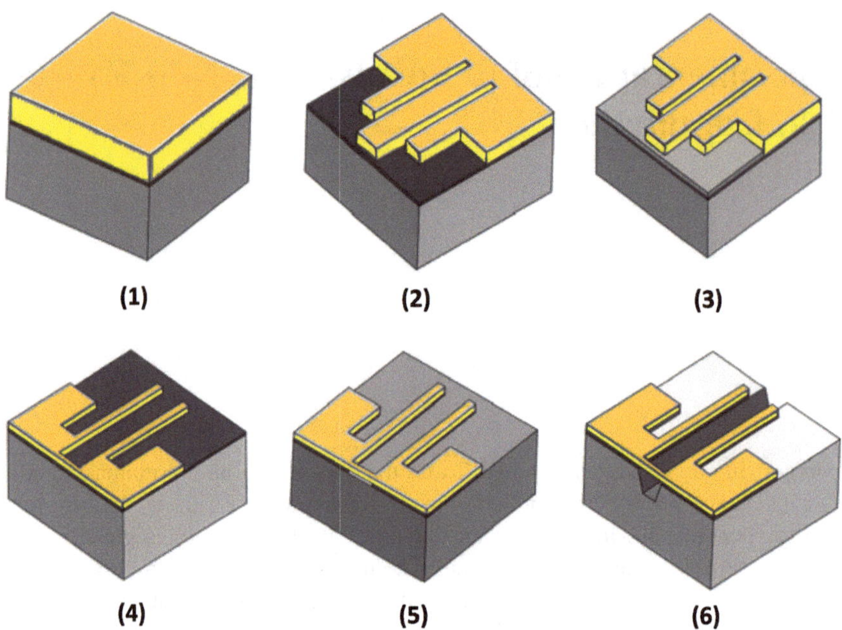

Fig. 4.1 Fabrication procedures of suspended metallic nanofilms [1]. The detailed information from step (1) to step (6) is given in the text

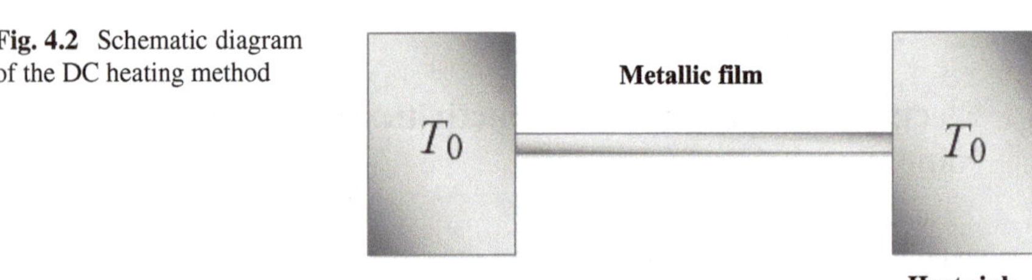

Fig. 4.2 Schematic diagram of the DC heating method

liquid resist-remover to leave only the metallic films on the substrate; (5) The isotropic etching is used to remove the SiO_2 layer around the nanofilm; the titanium layer is also etched away; (6) Si substrate is anisotropically etched using KOH solution, leaving the metallic nanofilm suspended above the substrate. The etched gap is about 6 μm in depth. As shown in the Fig. 4.1 (6), the suspended nanofilm is well connected with the leads on the both sides, no contact resistance exists. The free-standing nanofilm is designed to avoid heat leak into the substrate.

A direct current (DC) heating method is applied to measure the electrical and thermal conductivities of the metallic nanofilms. A schematic diagram of this method is shown in the Fig. 4.2.

In the Fig. 4.2, the fabricated metallic nanofilm is suspended above the substrate and connects with the leads on its both sides. The temperature of lead is the ambient temperature T. In the DC heating method, the metallic nanofilm serves as both Joule heater and thermometer. In a high vacuum environment, the heat convection can be ignored. On the other hand, assuming the temperature of nanofilm is 3 K, the heat radiation from 300 K environment gives 0.0037 μW to the film; the electrical heating power during the measurement is 23.36 μW, thus the heat radiation can be ignored as well. The DC heating causes a parabolic temperature distribution along the nanofilm and the change of the resistance is proportional to the average temperature rise of the film. The thermal conductivity of the nanofilm can be obtained by measuring the electrical power, resistance, and temperature coefficient.

Solving an one-dimensional heat conduction equation, the temperature distribution of the nanofilm can be obtained as

$$T(x) = T_0 + \frac{IU}{2w\delta\kappa_f}x - \frac{IU}{2w\delta l\kappa_f}x^2 \tag{4.1}$$

where x is the distance from the left end of the film; U and I are the voltage and current over the film; l, w, and δ are the length, width, and thickness of the film, respectively; κ_f is the thermal conductivity. In order to minimize the uncertainty induced by the temperature change, the electrical heating power is at microwatts level and the corresponding temperature rise is below 10 K. The average temperature rise calculated from the Eq. (4.1) is:

$$\Delta T_l = \frac{IUl}{12w\delta\kappa_f} \tag{4.2}$$

Meanwhile, the linear relationship between the temperature rise and change of resistance is

$$R_f = R_0 + \beta_f R_r \Delta T_l = R_0 + \frac{\beta_f R_r}{\kappa_f}\frac{IUl}{12w\delta} \tag{4.3}$$

where R_f, R, and R_r are the film resistance, resistances at T, and at reference temperature T_r, respectively. β_f is the temperature coefficient of the nanofilm, given as

$$\beta_f = \frac{R_0 - R_r}{R_r\,(T_0 - T_r)} \tag{4.4}$$

The resistance of the nanofilm is measured using a four-probe method.

Figure 4.3 is the screenshot of the original experimental data of 76 nm gold (Au) nanofilm. The red circles mark the linear fitting results of the resistance versus electrical power data, the R^2 value is more than 0.999. The thermal conductivity of the nanofilm can be calculated as

$$\kappa_f = \frac{\beta_f R_r}{b}\frac{l}{12w\delta} \tag{4.5}$$

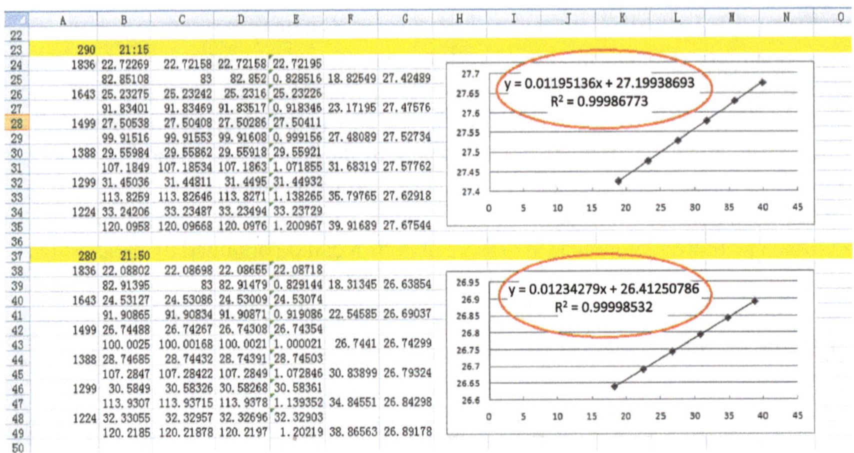

	A	B	C	D	E	F	G
23	290	21:15					
24	1836	22.72269	22.72158	22.72158	22.72195		
25		82.85108	83	82.852	0.828516	18.82549	27.42489
26	1643	25.23275	25.23242	25.2316	25.23226		
27		91.83401	91.83469	91.83517	0.918346	23.17195	27.47576
28	1499	27.50538	27.50408	27.50286	27.50411		
29		99.91516	99.91553	99.91608	0.999156	27.48089	27.52734
30	1388	29.55984	29.55862	29.55918	29.55921		
31		107.1849	107.18534	107.1863	1.071855	31.68319	27.57762
32	1299	31.45036	31.44811	31.4495	31.44932		
33		113.8259	113.82646	113.8271	1.138265	35.79765	27.62918
34	1224	33.24206	33.23487	33.23494	33.23729		
35		120.0958	120.09668	120.0976	1.200967	39.91689	27.67544
37	280	21:50					
38	1836	22.08802	22.08698	22.08655	22.08718		
39		82.91395	83	82.91479	0.829144	18.31345	26.63854
40	1643	24.53127	24.53086	24.53009	24.53074		
41		91.90865	91.90834	91.90871	0.919086	22.54585	26.69037
42	1499	26.74488	26.74267	26.74308	26.74354		
43		100.0025	100.00168	100.0021	1.000021	26.7441	26.74299
44	1388	28.74685	28.74432	28.74391	28.74503		
45		107.2847	107.28422	107.2849	1.072846	30.83899	26.79324
46	1299	30.5849	30.58326	30.58268	30.58361		
47		113.9307	113.93715	113.9378	1.139352	34.84551	26.84298
48	1224	32.33055	32.32957	32.32696	32.32903		
49		120.2185	120.21878	120.2197	1.20219	38.86563	26.89178

Fig. 4.3 Screenshot of the original measured data

where b is the slope in Fig. 4.3. If R is the resistance at zero electrical heating power, the electrical conductivity σ_f is:

$$\sigma_f = \frac{l}{R_0 w \delta} \tag{4.6}$$

DC heating method can be applied with high accuracy, especially for the metallic materials.

It has already been pointed out in Chap. 2 that the thermomass inertia plays an important role under the ultra-high heat flux and low temperature conditions. In the experiment, a liquid helium cooling system is used to provide the minimum 2.8 K environmental temperature.

Figure 4.4 shows the PT403 liquid helium cooling system of Cryomech Inc. Figure 4.4a shows the low temperature high vacuum chamber, a thermostatic platform is placed inside. A radiation shield is used to eliminate the heat radiation effect. The test film chip is attached to the thermostatic platform, whose temperature can be controlled from 2.8 to 300 K. Figure 4.4b is the liquid helium compressor to provide a closed loop cold source.

Figure 4.5 shows the ITC503 intelligent temperature controller from Oxford Inc. At high temperatures, the temperature resolution is 0.1 K; at low temperatures below 20 K, the temperature resolution is 0.01 K. In the experiment, at least 40 min should be waited for a stable temperature.

A four-probe method is used to measure the film resistance accurately. Figure 4.6 shows the used 2002 high precise voltmeters in the experiment, where a voltmeter and a 100 Ω standard resistance from Yokogawa Inc. are used to measure the current.

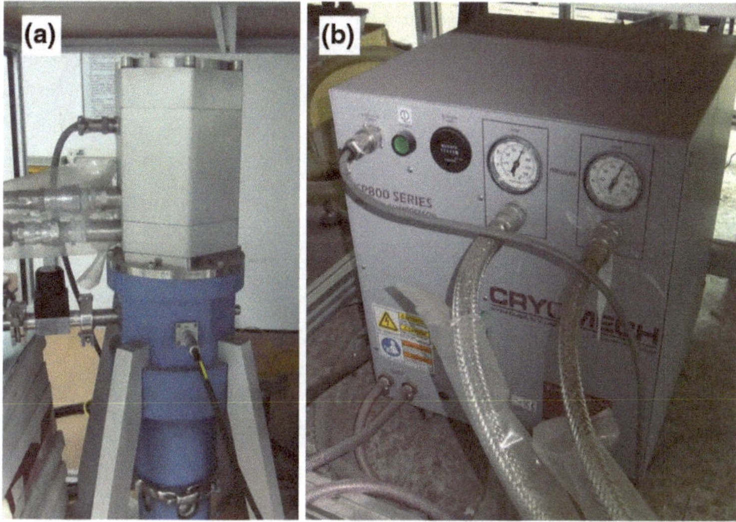

Fig. 4.4 **a** High vacuum chamber and **b** liquid helium compressor

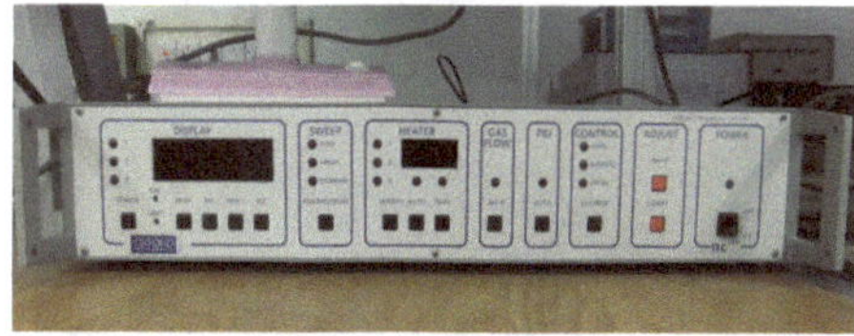

Fig. 4.5 Intelligent temperature controller

Fig. 4.6 High precise voltmeters

Figure 4.6 shows the high precise voltmeters, whose range is from 1 nV to 1,100 V. The above one is used to measure the voltage and the below one is used to measure the current.

Figure 4.7 shows the two-stage air pumping system formed by a mechanical pump and a molecular pump. The achieved air pressure can be below 10^{-4} Pa, thus the convection can be ignored.

Fig. 4.7 Two-stage air pumping system

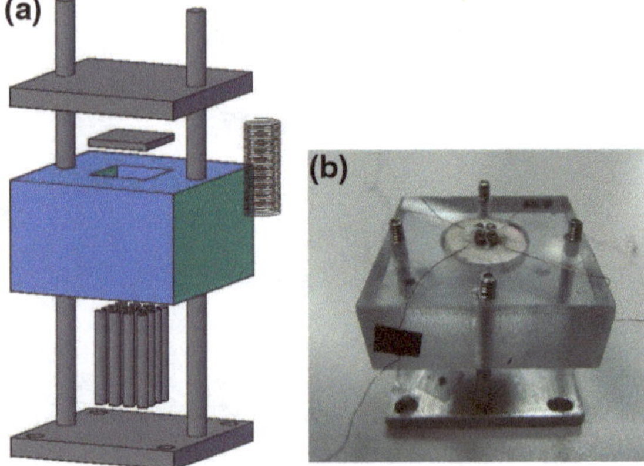

Fig. 4.8 Spring-pin contact device for electrical connection. **a** The assembly diagram of spring-pin contact device. **b** The photo of spring-pin contact device

There are two key problems need be solved to ensure high measurement accuracy:

1. Good electrical connection at very low temperatures. The commonly used welding method cannot be applied for the nanofilm. The silver paste is usually valid in the middle temperature range, losing its conductivity at very low temperatures (<80 K); the silver particles inside lose their activities. During the pasting process, the static electricity from the human body may break the nanofilm. In order to keep a good electrical connection between the nanofilm and the lead wires, a spring-pin contact device has been developed as shown in the Fig. 4.8.

Figure 4.8a shows the 3D design drawing of the contact device and Fig. 4.8b shows the photograph. Four 500 μm diameter copper wires are welded with the metal pins,

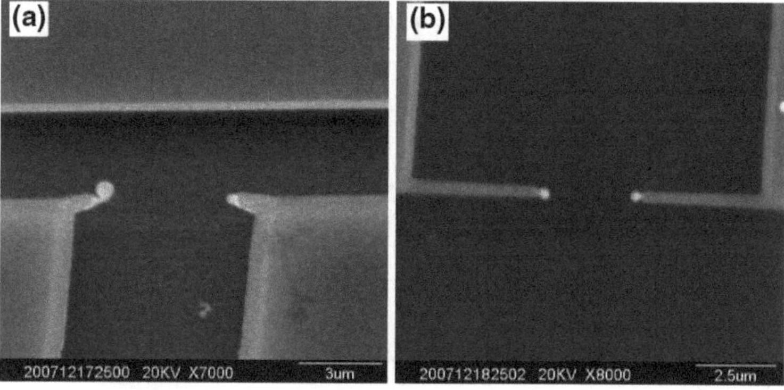

Fig. 4.9 Several broken metallic nanofilms. **a** An electrically broken nanofilm around 7.0 um. **b** An electrically broken nanofilm around 10.8 um

then the pins are pressed by four springs onto the Si chip to directly connect with the nanofilm. The contact resistance is measured to be below 80 mΩ. In this way, the good electrical connection can be ensured at minimum 2.8 K for more than 15 h. Meanwhile, two metal plates are placed above and below the Si chip to prevent heat radiation.

1. Avoid nanofilm breaking by electrical impulses. Because the metallic nanofilm is at a nanometer level, even small electrical impulses could break the nanofilm. The Fig. 4.9 shows several broken nanofilms.

It is seen in the Fig. 4.9 that the metallic nanofilms are broken in the middle, due to the overheating of transient electrical impulses. At daytime, the voltage fluctuation of the power grid will burn the nanofilms, especially at peak hours. In order to solve this problem, the experiments were arranged at night, usually from 18:00 pm to 9:30 am in the next morning. During the night time, the electrical voltage was more stable. Second, a protection circuit was designed to avoid electrical impulse damage. The transient impulse current could be conducted to the by-pass circuit.

4.1.2 Electrical Conductivity

The electrical conductivities of several Au nanofilms have been measured, and the results are analyzed using a transport theory to investigate the size effect in depth. When the film thickness is comparable with the mean free path (MFP) of electrons, the electron scattering at surfaces is the main contribution factor for the size effect. Fuchs [2] and Sondheimer [3] developed a prediction equation for the electrical conductivity of nanofilm based on Boltzmann transport equations:

$$\frac{\sigma_f}{\sigma_b} = 1 - \frac{3(1-p)}{2k} \int_1^\infty \left(\frac{1}{t^3} - \frac{1}{t^5}\right) \frac{1-\exp(-kt)}{1-p\exp(-kt)} dt \tag{4.7}$$

where σ_b and σ_f are the electrical conductivities of the bulk and nanofilm, respectively. p is the surface reflection coefficient and k is the ratio between the film thickness δ and MFP l_b. The FS theory considered the surface scattering of electrons to be responsible for the size effect. Some other researchers [4] pointed out that the electron scattering within one MFP from the surface was effective, because the MFP was highly related to the temperature and the surface scattering was also sensitive to the temperature. But when the MFP of electrons is much larger than the film thickness, or in the polycrystalline films, the surface scattering is not the only dominant factor any more.

Mayadas and Shatzkes [5, 6] studied the size effect of electrical conductivity for the polycrystalline films and found that the grain boundary scattering was significant. They introduced a δ potential function into the Boltzmann transport equation and the prediction equation was derived as

$$\frac{\sigma_f}{\sigma_b} = 1 - \frac{3}{2}\alpha + 3\alpha^2 - 3\alpha^3 \ln\left(1 + \frac{1}{\alpha}\right) \tag{4.8}$$

where $\alpha = l_b R/d(1-R)$ is a combined parameter, including average grain size d, MFP of electrons l_b, reflection coefficient at grain boundaries R, l_b, and R are unknown parameters in the experiment. The MFP can be calculated as the product of the relaxation time and Fermi speed: $l_b = \tau \times v_F$. Based on the classical kinetic theory of electrons, the electrical conductivity and MFP follow the Drude's relation [3]:

$$\sigma_b = \frac{ne^2\tau}{m} = \frac{ne^2 l_b}{m v_F} \tag{4.9}$$

where n, e, and m are number density, elementary charge, and mass of electron. The bulk value can be calculated using the Bloch–Grüneisen model as [7]:

$$\rho_b = \rho_0 + \rho_{ee}T^2 + \rho_{sd}T^3 \frac{J_3}{7.212} + \rho_{ss}T^5 \frac{J_5}{124.14}, \tag{4.10}$$

where

$$J_N = \int_0^{\Theta/T} \frac{x^N dx}{(e^x - 1)\left(1 - e^{-x}\right)} \tag{4.11}$$

In the Eq. (4.10), the four terms on the right side of equal sign are the zero resistance, electron–electron scattering, interband, and intraband electron scattering, respectively. The coefficients of these four terms can be only decided by the experiments.

The film samples made of 99.98 % purity metal are provided by Prof. Takahashi from Kyushu University. The fabricated films are shown in the Fig. 4.10.

Fig. 4.10 Scanning electron microscope (SEM) image of the fabricated film sample. **a** SEM image of nanofilm from a side view. **b** SEM image of nanofilm from a top view. Reprinted from "Non-Fourier heat conduction study for steady states in metallic nanofilms, 57, Hai-Dong Wang et al.". Copyright [2012], with permission from Springer

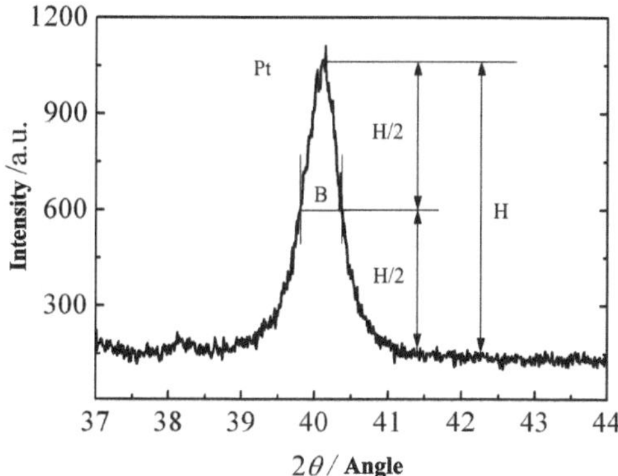

Fig. 4.11 X-ray diffraction measurement result. Reprinted from "Experimental study on the influences of grain boundary scattering on the charge and heat transport in gold and platinum nanofilms, 47, Hai-Dong Wang et al.". Copyright [2011], with permission from Springer

Figure 4.10 shows the SEM images of Au nanofilms, (a) and (b) are the images from different view angles. The average grain size can be measured using X-ray diffraction method as shown in the Fig. 4.11.

According to the measured data in Fig. 4.11, the average grain sizes are: $d = 15.3$ nm for 48 nm thick platinum (Pt) film; $d = 38.3$ nm for 53 nm thick Au film; and $d = 45.3$ nm for 76 nm thick Au film.

Figure 4.12 shows the measured electrical conductivity of 48 nm thick Pt film, where the inset is the calculated MFP of electrons. The MFP increases as the temperature decreases, reaching tens of micrometers below 50 K. At low temperatures, the MFP is much larger than the grain size, the electron scattering contributes greatly to the size effect.

Figure 4.13 shows the measured results of four Au nanofilms. As shown in the Figs. 4.12 and 4.13, the measured data of nanofilms are greatly reduced from the bulk value due to the size effect. The size effect increases as the temperature decreases. In the Fig. 4.13, all the film samples have been annealed in the vacuum for more than half an hour. The heating current is about 6.32 mA, causing 220 K average

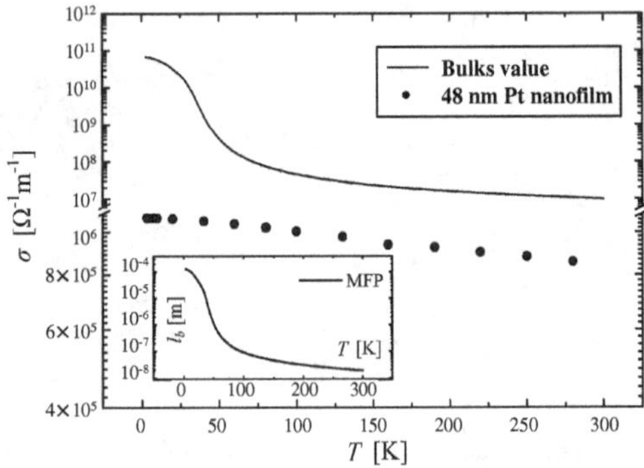

Fig. 4.12 Measured electrical conductivity of 48 nm thick Pt film. Reprinted from "Experimental study on the influences of grain boundary scattering on the charge and heat transport in gold and platinum nanofilms, 47, Hai-Dong Wang et al.". Copyright [2011], with permission from Springer

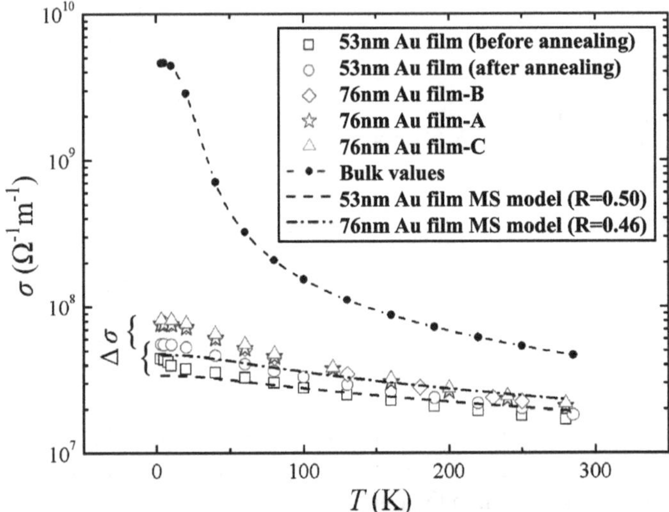

Fig. 4.13 Electrical conductivities of Au nanofilms with different thicknesses. Reprinted from "Breakdown of Wiedemann-Franz law in individual suspended polycrystalline gold nanofilms down to 3K, 66, Haidong Wang, Jinhui Liu, Xing Zhang, Koji Takahashi, 585–591." Copyright [2013], with permission from Elsevier

temperature rise. In order to examine the annealing effect on the nanostructures, the same 53 nm thick nanofilm has been measured twice before and after annealing process. Some defects and dislocations in the nanofilm are removed, thus the electrical conductivity can be improved.

The electrical conductivity is predicted using the MS model combined with Drude's relation. At high temperatures, the predicted result agrees well with the measured data. The fitted reflection coefficients are determined to be 0.50 and 0.46

for 53 and 76 nm films, respectively. At temperature below 100 K, the prediction of MS model is significantly lower than the measured data and the deviation $\Delta\sigma/\sigma_f$ could be more than 40 %.

At high temperatures, the MS model can be successfully applied to predict the electrical conductivity and points out that the grain boundary scattering becomes the dominant factor since the MFP of electrons is comparable with the grain size. But an obvious deviation occurs with a decreasing temperature, this is mainly due to the breakdown of Dude's relation. There are three assumptions for Drude's relation: (1) the reflection coefficient R is independent to temperature; (2) average grain size d is independent to temperature; (3) the collisions between electrons are elastic and the electrical conductivity is proportional to the bulk MFP. The previous assumptions (1) and (2) are known to be valid for nanofilms [8], but the third assumption breaks at low temperatures and the MS model loses its accuracy.

For monocrystalline metallic films, the background scattering of electrons is the same as that in bulks, thus the elastic Drude's relation is valid. But for polycrystalline films, the grain boundary scattering plays an important role. At high temperatures, the MFP of electrons is much smaller than the grain size and the limitation of grain boundary scattering is not significant, thus the background scattering is still elastic like the case in monocrystalline films. At low temperatures, the MFP can be much larger than the grain size, thus the grain boundary scattering becomes the only dominant factor. Some electrons are inelastically scattered at the grain boundaries, part of the electron energy is transferred to the phonons across the boundaries. In this way, the elastic Drude's relation breaks, and the MFP is decreased. In order to modify the MS model, an effective MFP of electrons is introduced, that is the MFP considering the inelastic grain boundary scattering. The reflection coefficient R at grain boundaries is temperature independent and can be decided at high temperatures when the Drude's relation is valid. Then the effective MFP l_b' becomes the only unknown parameter related to the electrical conductivity.

Figure 4.14 shows the comparison between the effective MFP l_b' and elastic MFP l_b from the Drude's relation. At high temperatures (>80 K), l_b' equals l_b. Then the difference between l_b' and l_b increases as the temperature decreases. Below 20 K, l_b' is significantly lower than l_b and the inelastic scattering plays an important role.

4.1.3 Thermal Conductivity

A significant size effect is also observed in the thermal conductivity of nanofilms. But there are only a few methods available for theoretical predictions, normally the thermal–electrical analogy method is used to calculate the thermal conductivity from the measured electrical conductivity. Tien [9] and Qiu [10] studied the size effect of heat conduction in nanofilms. Based on the thermal–electrical analogy, the surface scattering mechanism was analyzed using the FS model; the grain boundary scattering mechanism was analyzed using the MS model. Kumar [11] studied the thermal conductivity of monocrystalline nanofilms using Boltzmann transport equation, the

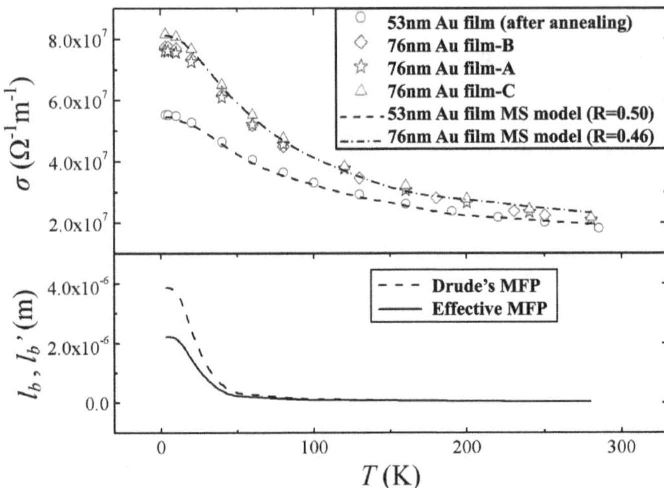

Fig. 4.14 Effective MFP of electrons extracted from the experimental data. Reprinted from "Breakdown of Wiedemann-Franz law in individual suspended polycrystalline gold nanofilms down to 3K, 66, Haidong Wang, Jinhui Liu, Xing Zhang, Koji Takahashi, 585–591." Copyright [2013], with permission from Elsevier

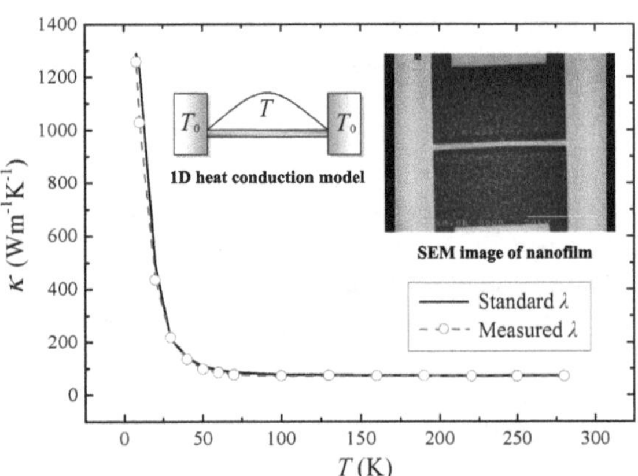

Fig. 4.15 Comparison between the measured thermal conductivity and standard value of Pt wire. Reprinted from "Breakdown of Wiedemann-Franz law in individual suspended polycrystalline gold nanofilms down to 3K, 66, Haidong Wang, Jinhui Liu, Xing Zhang, Koji Takahashi, 585–591." Copyright [2013], with permission from Elsevier

obtained prediction formula was much like the one of electrical conductivity. In order to investigate the effects of different electron scattering mechanisms on thermal conductivity, several metallic nanofilms are prepared and measured in the experiment.

The thermal conductivity of a 30 μm diameter Pt wire has been measured to compare with the standard value as shown in the Fig. 4.15.

The red circles in the Fig. 4.15 are the measured data and the black solid curve is the standard value. The good agreement shows high measurement accuracy. The insets are the SEM image of Au nanofilm and one dimensional heat conduction model of Pt wire. When the nanofilm is heated by a direct current, a parabolic temperature profile will exist along the film.

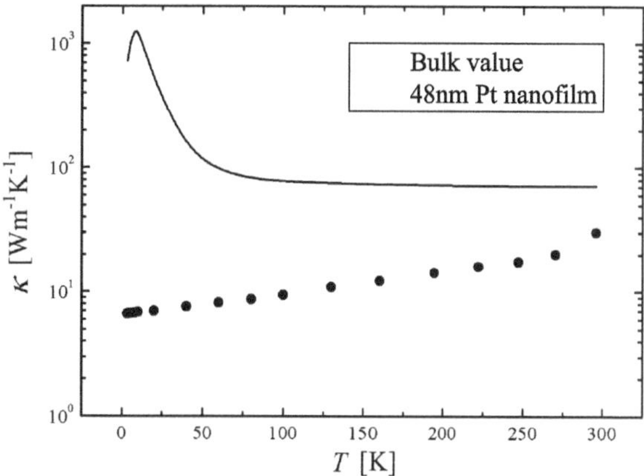

Fig. 4.16 Thermal conductivity of 48 nm thick Pt nanofilm

Figure 4.16 shows the measured thermal conductivity of Pt nanofilm plotted with respect to temperature. It is found that the thermal conductivity decreases as the temperature decreases, while an opposite temperature trend is observed for bulk values. This deviation is due to the significant size effect caused by several electron scattering mechanisms in nanofilms. Matthiessen's rule can be applied to calculate different scattering mechanisms quantitatively and is widely used to predict the thermal conductivities of dielectrics [12] and metallic nanostructures [13]. Matthiessen's rule can be written as

$$\frac{1}{\tau_{\text{film}}} = \frac{1}{\tau_{\text{background}}} + \frac{1}{\tau_{\text{surface}}} + \frac{1}{\tau_{\text{grain}}} + \frac{1}{\tau_{\text{impurity}}} \tag{4.12}$$

The four terms on the right side of equal sign are: electron background scattering, surface scattering, grain boundary scattering, and impurity scattering, respectively. Because the nanofilms are made from high-purity metals, the impurity scattering can be ignored. For polycrystalline films, the grain boundary scattering dominates over the surface scattering, and the latter one can be ignored. Thus the Eq. (4.12) can be simplified as:

$$\frac{1}{\tau_{\text{film}}} = \frac{1}{\tau_{\text{background}}} + \frac{1}{\tau_{\text{grain}}} = \frac{v_F}{l_b} + \frac{v_F}{d} \tag{4.13}$$

Both the background scattering and grain boundary scattering should be taken into consideration. R is the percentage of electrons reflected from the grain boundaries, taking R into account, the Eq. (4.13) can be written as:

$$\frac{1}{\tau_{\text{film}}} = \frac{1}{\tau_{\text{background}}} + \frac{1}{\tau_{\text{grain}}} = \frac{v_F}{l_b} + R\frac{v_F}{d}. \tag{4.14}$$

As a result, the ratio between the thermal conductivities of nanofilm and bulk is

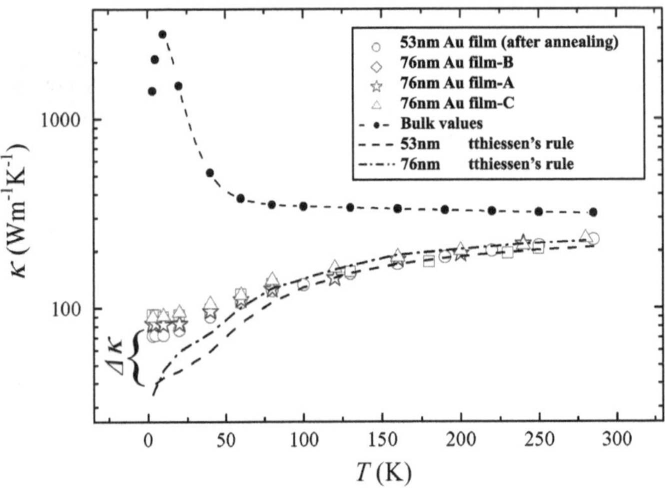

Fig. 4.17 Thermal conductivities of Au nanofilms with different thicknesses. Reprinted from "Breakdown of Wiedemann-Franz law in individual suspended polycrystalline gold nanofilms down to 3K, 66, Haidong Wang, Jinhui Liu, Xing Zhang, Koji Takahashi, 585–591." Copyright [2013], with permission from Elsevier

$$\frac{\kappa_f}{\kappa_b} = \frac{\tau_{\text{film}}}{\tau_{\text{background}}} = \frac{v_F}{l_b} \Big/ \left(\frac{v_F}{l_b} + R\frac{v_F}{d}\right) = \frac{1}{1 + Rl_b/d} \tag{4.15}$$

If there is no grain boundary scattering, $R = 0$ and κ_f equals κ_b.

Figure 4.17 shows the measured thermal conductivities of Au nanofilms, the prediction result based on the Matthiessen's rule is also plotted for comparison. The measured data are much smaller than the corresponding bulk values due to the significant size effect. Using the elastic MFP of electrons, the prediction of Matthiessen's rule agrees well with the measured data at high temperatures. But below 80 K, the theoretical prediction is obviously lower than the experimental result and the relative deviation $\Delta\kappa/\kappa_f$ could be more than 50 %. Similarly, the inelastic electron scattering at grain boundaries is responsible for the failure of Matthiessen's rule. Using the effective MFP instead, the predicted thermal conductivity is plotted in the Fig. 4.18.

Figure 4.18 shows the theoretical prediction of Matthiessen's rule using the effective MFP instead. The theoretical prediction agrees well with the experimental results. It should be noted that the used effective MFP and reflection coefficient R are the same as those used for electrical conductivity. It has been mentioned above that the inelastic scattering will decrease the MFP. This is due to the increased electron–phonon collisions inside the grains. Part of the electron energy is transferred across the grain boundaries and carried on by phonons, thus the phonon thermal conductivity is increased accordingly. This additional thermal conductivity is neglected in the Fig. 4.17, so the theoretical predictions are lower than the measured data at low temperatures.

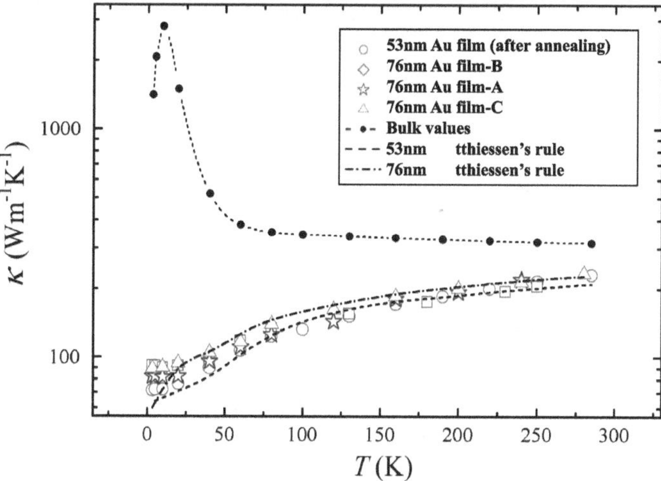

Fig. 4.18 Thermal conductivity calculated using effective MFP. Reprinted from "Breakdown of Wiedemann-Franz law in individual suspended polycrystalline gold nanofilms down to 3K, 66, Haidong Wang, Jinhui Liu, Xing Zhang, Koji Takahashi, 585-591." Copyright [2013], with permission from Elsevier

Stojanovic [14] studied the increased phonon thermal conductivity in nanomaterials using Boltzmann transport equation. The contribution of phonons to the thermal conductivity increases as the material size decreases. But Stojanovic's research is still limited at room temperatures. At low temperatures, the increased phonon thermal conductivity will cause breakdown of Wiedeman–Franz law.

4.1.4 Break Down of Wiedemann–Franz Law at Low Temperatures

Based on quantum mechanics theory, Smekal [15], Kramers and Heisenberg [16] Schroedinger [17] and Dirac [18] predicted the existence of an inelastic light scattering phenomenon. Later in 1928, Raman observed this phenomenon in liquid for the first time, named as Raman scattering [19] At the same year, Landsberg and Mandelstam [20] observed the same phenomenon in crystals. From a microscopic point of view, the incident photons with energy hw_I and momentum k_I are scattered and the scattered photons have energy hw_S and momentum k_S. During this process, an elementary excitation with energy hw and momentum q will be created or annihilated. The energy and momentum conservation equations are listed as:

$$\hbar w_S = \hbar w_I \pm \hbar w \tag{4.16}$$

$$k_S = k_I \pm q \tag{4.17}$$

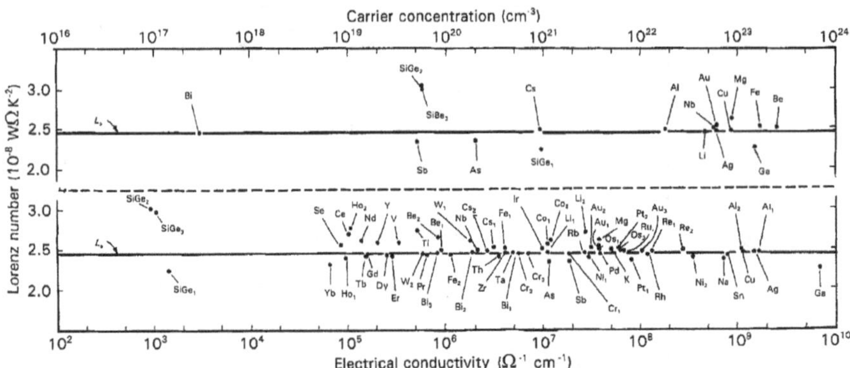

Fig. 4.19 Measured Lorenz number of metals [25]. Reprinted with the permission from Ref. [25]. Copyright [1993], with permission from Elsevier

where the positive and negative signs stand for the Stokes and anti-Stokes processes, respectively. The physical essence of Raman scattering is the energy and momentum exchange between the incident particles and materials. During the Stokes process, the photons lose energy hw; while during the anti-Stokes process, the photons gain energy hw. On the other hand, if $hw_I = hw_S$, the scattering is elastic.

Similarly, the electrons can be scattered elastically and inelastically, the energy is not conserved during inelastic scattering, and it is referred to as the electron Raman scattering. For photons, the elastic scattering dominates in the most cases, and the probability of Raman scattering is only about 1/1,000. For electrons, the elastic scattering is also dominant, this point of view can be proved by the fact that the Wiedeman–Franz (WF) law holds robustly. The WF law is normally seen as one of the great achievements of classical electron theory. It states that the ratio of charge and heat transport by electrons is proportional to temperature. In another way of speaking, $\kappa/\sigma T$ is a constant, which is referred to as Lorenz number [21]. Later, Sommerfeld [22] obtained the exact value of Lorenz number as $L_0 = \kappa/\sigma T = (\pi k_B/e)^2/3 = 2.443 \times 10^{-8}\,\mathrm{W}\Omega\mathrm{K}^{-2}$, where k_B is Boltzmann constant, e is the elementary charge. The WF law has been proved to be true for the most metals, no matter at high or low temperatures [23, 24].

Figure 4.19 summarized the measured Lorenz number of different metals plotted with respect to carrier concentration or electrical conductivity. The solid line is the theoretical value of L. It is seen that the most results agree well with L, thus the WF law is widely accepted. But some researchers pointed out that the Lorenz number did not always equal to L, except for the highly degenerated electron gas. The classical electron gas model ignores the electron–electron and electron-lattice interactions, the electron collisions are elastic.

Figure 4.19 proves the validation of WF law for bulk metals, but is it valid for nanomaterials? By solving Boltzmann transport equations, Ouarbya [26] proved that the WF law was valid for nanomaterials as long as the electron scattering was elastic. Meanwhile, Ziman [12] proved that the assumption of elastic scattering was valid for

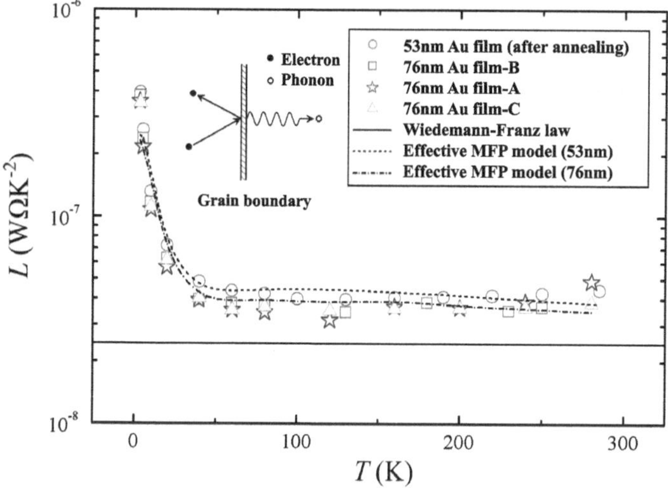

Fig. 4.20 Measured Lorenz number of Au nanofilms plotted with respect to temperature. Reprinted from "Breakdown of Wiedemann-Franz law in individual suspended polycrystalline gold nanofilms down to 3K, 66, Haidong Wang, Jinhui Liu, Xing Zhang, Koji Takahashi, 585–591." Copyright [2013], with permission from Elsevier

the most bulk metals. So the reason of breakdown of the WF law in nanomaterials may come from the inelastic electron scattering (electron Raman scattering). At present, there are no fully developed theories for the inelastic scattering of electrons; normally the electron–phonon and electron-impurity scatterings play an important role at low temperatures. In our case, the electron scattering at grain boundaries plays a dominant role.

In the Fig. 4.20, an increasing Lorenz number with decreasing temperature is observed, revealing that the WF law breaks in nanofilms at low temperatures. Above 50 K, the measured Lorenz number is larger than L and keeps a constant; below 50 K, the Lorenz number increases as the temperature decreases. The physical mechanisms are explained as follows:

1. At high temperatures (50 ~ 300 K), the electron Raman scattering at grain boundaries is not significant. When the electrons are scattered at grain boundaries, the charge transport is ended, while part of electron energy is transported by lattice vibrations. In this way, the effective thermal conductivity is relatively higher, causing a larger Lorenz number.
2. At low temperatures (<50 K), the thermal–electrical analogy breaks and the electron Raman scattering plays an important role.

Figure 4.21 clearly shows that the Au film samples are polycrystalline. Many perpendicular grain boundaries exist in the film as shown in the Fig. 4.22, different colors are used to mark the crystal grains. The electrons have to travel through all the grain boundaries to transport charge and heat.

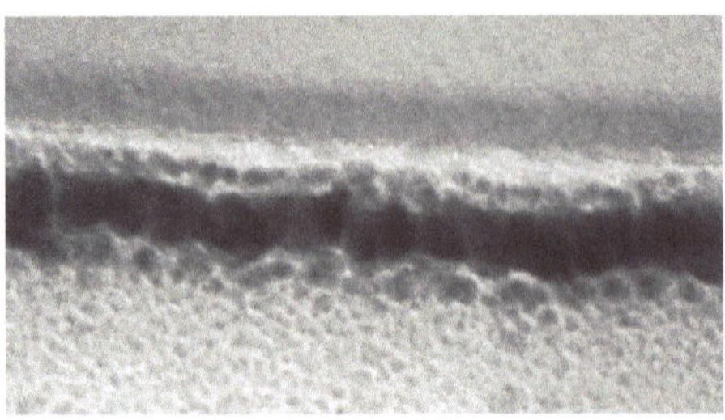

Fig. 4.21 Transmission electron microscope (TEM) image of the cross-sectional area of Au nanofilm

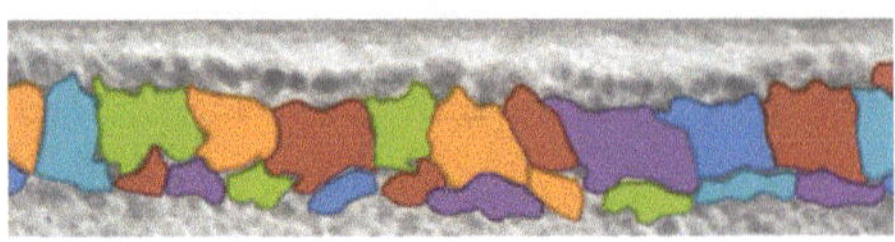

Fig. 4.22 Grain boundaries plotted on the TEM image

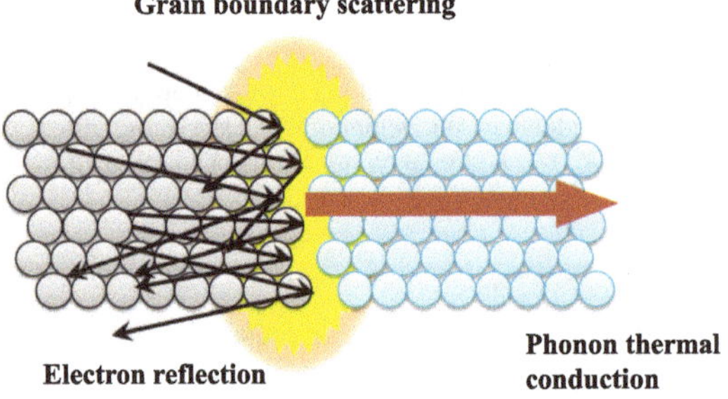

Fig. 4.23 Electron Raman scattering at grain boundaries

Figure 4.23 is the schematic diagram of electron Raman scattering process. At low temperatures, the MFP of electrons is much larger than the grain size and the electron-lattice scattering becomes the only dominant factor. During the electron Raman scattering, part of electron energy is transported to the lattices and carried on by lattice vibration. The phonons are the quantum mechanical particles within the lattice waves, the phonon thermal conductivity will be increased during electron Raman scattering process. On the other hand, the electron Raman scattering has no effect on the charge transport. Hence, the Lorenz number is increased and the WF law breaks.

The effective MFP of electrons can be used to predict the electrical and thermal conductivities theoretically. Combining the MS model and Matthiessen's rule together, the Lorenz number can be well predicted [27, 28]:

$$L_f = \frac{\lambda_f}{\sigma_f T} = \frac{L_b}{\left(1 + Rl_b'/d\right)\left[1 - \frac{3}{2}\alpha + 3\alpha^2 - 3\alpha^3 \ln\left(1 + \frac{1}{\alpha}\right)\right]} \tag{4.18}$$

where $L_b = \kappa_b/\sigma_b T$ is the Lorenz number of bulk materials and l_b' is the effective MFP. The theoretical prediction based on the Eq. (4.18) is shown in the Fig. 4.20, matching well with the experimental data in a wide temperature range. The Eq. (4.18) can be applied to replace the WF law in calculating the thermal conductivity from the measured electrical conductivity.

4.2 Experimental Proof of Steady Non-Fourier Heat Conduction

The maximum heat flux in metallic nanofilms could exceed 10^{10} Wm^{-2} at low temperatures, satisfying the two conditions for non-Fourier heat conduction. In this chapter, the steady non-Fourier heat conduction is observed and studied in depth, providing experimental evidence for the thermomass theory.

4.2.1 Experimental Principle

One dimensional general heat conduction equation with internal heat source in steady states is written as:

$$q + \kappa_I \left(\frac{1 - \frac{q^2}{\frac{5}{2}\gamma_h \pi^2 n^3 k_B^3 T^3 / \rho}}{1 + \frac{2S\kappa_I}{\frac{5}{2}\gamma_h \pi^2 n^3 k_B^3 T^2 / \rho}} \right) \frac{dT}{dx} = 0 \tag{4.19}$$

In the Eq. (4.19), the term in parentheses is induced by spatial thermmass inertia. γ_h is the correction factor of the state equation of thermomass caused by the internal energy of lattices, it can be set as the Grüneisen parameter. The intrinsic thermal conductivity κ_I has already been measured using the DC heating method, where the electrical power is small enough to ensure a negligible thermomass inertia. Under the high heat flux and low temperature conditions, the temperature predicted by the general Eq. (4.19) is larger than that predicted by Fourier's law. In the experiment, the temperature rise of nanofilm is gained by measuring the film resistance. Compare the experimental data with the theoretical predictions, the validation of the general heat conduction law can be verified.

4.2.2 Experimental Result and Analysis

Before measurement, all the metallic nanofilms have been heated and annealed in vacuum. The annealing process will remove the defects in the film sample and decrease the resistance accordingly. In order to testify the effect of annealing process, the resistance of nanofilm has been measured twice.

Table 4.1 First measurement at 5 K after annealing

Power (μW)	10.6166	13.3313	15.8537	18.3723	20.8364	23.3564
Resistance (Ω)	6.0910	6.1526	6.2139	6.2778	6.3425	6.4097

Table 4.2 Second measurement at 5 K after annealing

Power (μW)	10.7594	13.2830	15.7968	18.3070	20.7614	23.2742
Resistance (Ω)	6.0932	6.1505	6.2115	6.2752	6.3393	6.4066

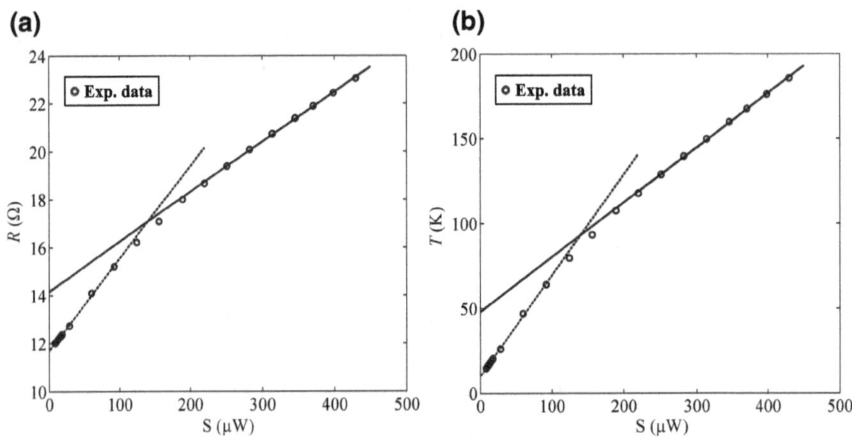

Fig. 4.24 Resistance and average temperature of the nanofilm plotted with respect to the electrical power. **a** Film resistance vs. electrical power. **b** Average temperature vs. electrical power

Tables 4.1 and 4.2 show the measured resistances of 76 nm thick Au nanofilm after the annealing process. Before the second measurement, the Au nanofilm was heated again by 6.32 mA current for half an hour. At the first time, the temperature coefficient of resistance (TCR) is 0.002571 K^{-1} and the thermal conductivity is 82.3 $Wm^{-1}K^{-1}$; at the second time, the TCR is 0.002572 K^{-1} and the thermal conductivity is 82.2 $Wm^{-1}K^{-1}$. The measured result demonstrates that the properties of the Au nanofilm reach a stable level. The measurement process of non-Fourier heat conduction causes no change to the properties of Au film sample.

When the nanofilm is heated by a large current in vacuum, a noticeable temperature rise could occur at the end of the film. Thus the temperature at the end of the film is $T + \Delta T$. The temperature rise ΔT could be evaluated from the change of the film resistance with respect to the electrical power.

Figure 4.24a gives the measured result of resistance changed with electrical power, and the Fig. 4.24b is the calculated average temperature from the resistance. When the electrical power is low, the intercept of the linearly fitted experimental data (broken line) is the environmental temperature T. For example, the intercept of the broken line in the Fig. 4.24b is 9.986 K, while $T = 10$ K in the experiment. When

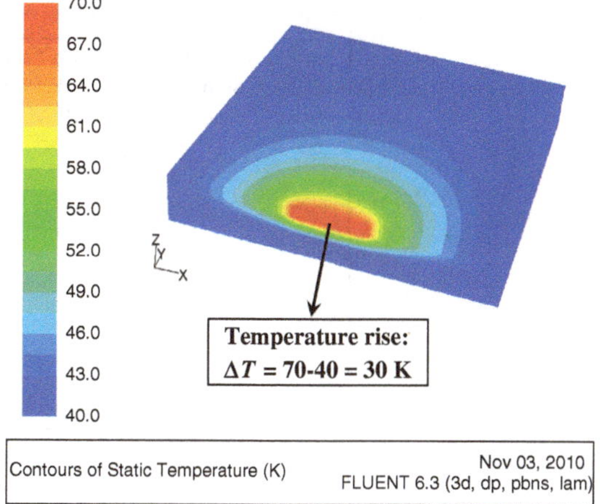

Fig. 4.25 3D calculation results using fluent software

the electrical power is larger, there will be a significant temperature rise ΔT at the end, thus the intercept of the linearly fitted data is $T + \Delta T$. The intercept of the solid line in the Fig. 4.24b is 48.082 K, so the temperature rise ΔT is evaluated to be $48.082 - 9.986 = 38.096$ K. As a matter of fact, the thermal conductivity increases as the temperature increases for nanofilms, so the slope of the *TS* line will be decreased and the intercept will be increased. In this way, the evaluated ΔT is actually larger than the exact value.

A three-dimensional heat conduction model is used to calculate the temperature rise at the end of nanofilm. A constant heat flux is used as the boundary condition and the other boundary temperatures are decided to be T. The thermal conductivity is set to be the measured data. The calculation result is shown in the Fig. 4.24.

Figure 4.25 is the calculation result obtained using Fluent software. The temperature rise is calculated to be 30 K, which is smaller than 38 K evaluated from the Fig. 4.24b. The temperature rise is taken into account in the theoretical analysis of non-Fourier heat conduction. The effects of heat flux and environmental temperature on the non-Fourier heat conduction are studied separately here.

(1) The effect of heat flux

In the experiment, three 76 nm thick Au films are measured to observe the non-Fourier heat conduction behavior. The Fig. 4.26 shows the measured thermal and electrical conductivities of the test films.

It is seen in the Fig. 4.26 that the experimental results of different Au nanofilms coincide with each other. The Au nanofilms with 302 and 308 nm in width are used to testify the validation of the general heat conduction law.

Figure 4.27 shows the measured average temperatures of different Au nanofilms at low temperatures, where the red circles are the experimental data, black squares are the prediction based on Fourier's law, blue stars are the prediction based on the general heat conduction law. It is noted that the predicted average temperature based on Fourier's law T_F is obviously lower than the experimental temperature

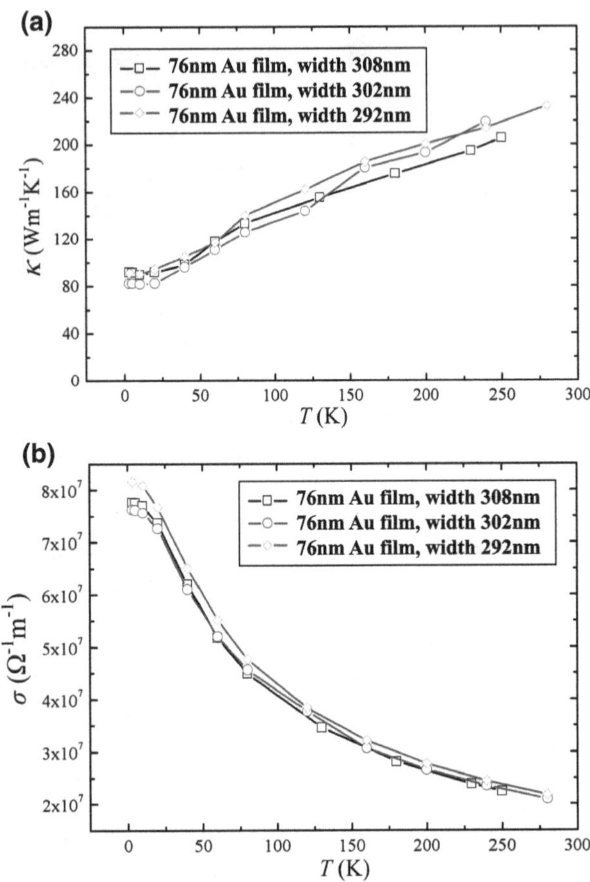

Fig. 4.26 Thermal and electrical conductivities of 76 nm thick Au films. **a** Temperature dependent thermal conductivity. **b** Temperature dependent electrical conductivity. Reprinted from "Non-Fourier heat conduction study for steady states in metallic nanofilms, 57, Hai-Dong Wang et al.". Copyright [2012], with permission from Springer

T_E, the difference $\Delta T = T_E T_F$ increases as the electrical power increases. The maximum heat flux in the experiment reaches $2.13\times10^{10}\,\text{Wm}^{-2}$ and the maximum average temperature is 284.50 K. The maximum $\Delta T_{\max}$ is 28.86 K. In order to see the temperature difference ΔT more clearly, zoom-in pictures of $\Delta T_{\max}$ are added as insets in the Fig. 4.27.

It is worthy of mentioning that the predicted average temperature based on the general heat conduction law T_G agrees well with the experimental temperature T_E. At low heat flux, T_F matches with T_E implying that the Fourier's law is still valid. In this case, the thermomass inertia effect is negligible and the general heat conduction law reduces to the Fourier's law. At high heat flux, the thermomass inertia is no longer

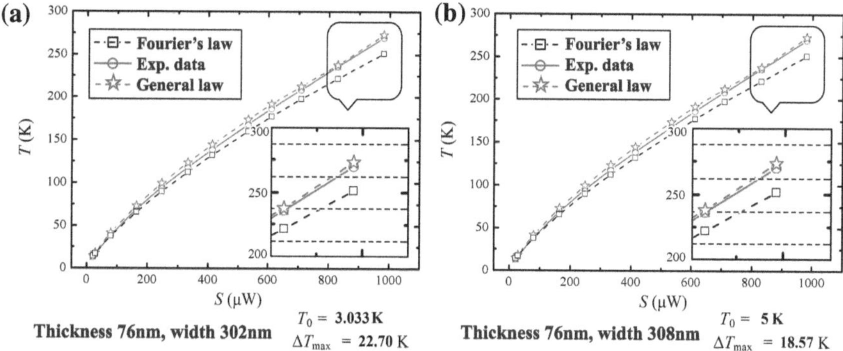

Fig. 4.27 Average temperature of the Au nanofilm plotted with respect to the electrical heating power. **a** Average temperature vs. electrical heating power at 3 K. **b** Average temperature vs. electrical heating power at 5 K. Reprinted from "Non-Fourier heat conduction study for steady states in metallic nanofilms, 57, Hai-Dong Wang et al.". Copyright [2012], with permission from Springer

negligible and the Fourier's law breaks. In this case, the general heat conduction law should be used instead.

(2) The effect of environmental temperature
The general heat conduction Eq. (4.19) states that the thermomass inertia effect increases as the heat flux increases and the environmental temperature decreases. In order to investigate the influence caused by the environmental temperature T, the same Au nanofilm has been continuously measured at different T from 18:00 pm to 9:30 am in the next morning. The film thickness, width, and length are 76, 292, and 9.51 μm, respectively. The experimental results are summarized in the Fig. 4.28.

The measured average temperatures of the Au nanofilm at different environmental temperatures are summarized in the Fig. 4.28. T is decreased from 60 to 3 K. The insets are the zoom-in pictures of $\Delta T_{\max} = T_E - T_F$. It is clearly seen that $\Delta T_{\max}$ increases from 6.81 to 23.62 K as T decreases. Meanwhile, the experimental data can be well predicted using the general heat conduction Eq. (4.19), proving the validation of the thermomass theory. But there is still a small deviation between T_E and T_G in the middle electrical power range, a good agreement is achieved under the low and high electrical power conditions. It is very likely that the deviation is caused by the uncertainty of the thermomass state equation, because: (1) at low electrical powers the thermomass inertia is negligible and the general heat conduction law reduces to the Fourier's law, thus the uncertainty of state equation has no effect; (2) at high electrical powers, the state equations in different materials come to a unified expression, the uncertainty induced by different microdistribution functions is decreased. At middle electrical powers, the uncertainty of state equation is relatively higher and becomes notable.

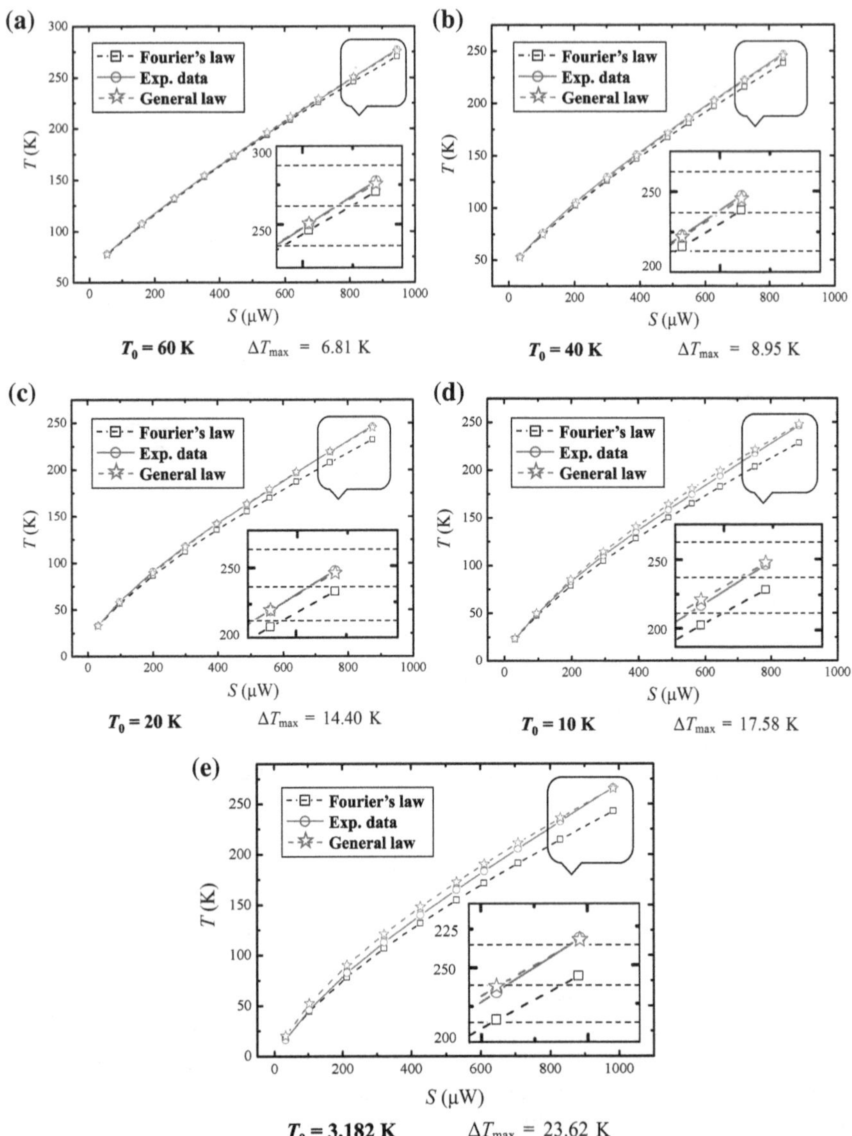

Fig. 4.28 Measured average temperatures of the Au nanofilm at different environmental temperatures: **a** 60 K, **b** 40 K, **c** 20 K, **d** 10 K, **e** 3.182 K

Table 4.3 shows the comparison between T_E, T_G and T_F.

It is noted that the measurements at different environmental temperatures are carried out at the same heat flux level, and the effect of heat flux is almost the same. The increasing $\Delta T_{\max}$ is mainly caused by the decreasing environmental temperature.

Table 4.3 Experimental data of non-Fourier heat conduction under high heat flux conditions

T (K)	Power (μW)	Heat-flux (Wm^{-2})	T_F (K)	T_E (K)	T_G (K)	$\Delta T_{\max}$ (K)
3	983.64	2.22×10^{10}	243.11	**266.42**	**265.57**	23.62
10	885.59	2.00×10^{10}	228.68	**246.19**	**246.88**	17.51
20	875.27	1.97×10^{10}	232.30	**246.61**	**245.16**	14.31
40	843.88	1.90×10^{10}	238.53	**247.37**	**245.43**	8.84
60	946.63	2.13×10^{10}	270.56	**277.00**	**275.70**	6.81

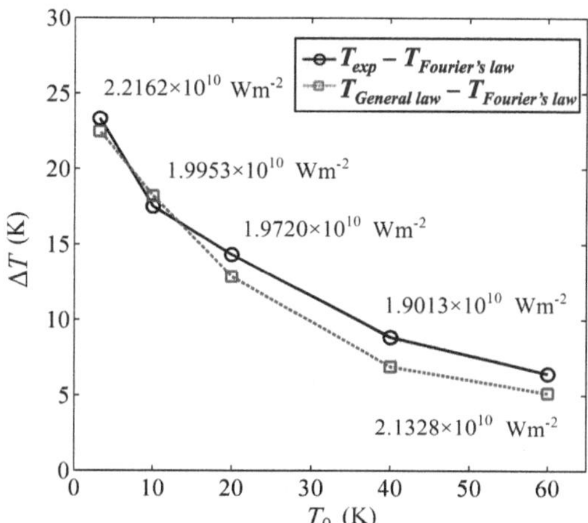

Fig. 4.29 $\Delta T_{\max}$ plotted with respect to the environmental temperature

Figure 4.29 shows the change of $\Delta T_{\max}$ with respect to the environmental temperature, the maximum heat flux is written in the figure. It demonstrates that the effect of thermomass inertia is enhanced with a decreasing temp erature at the same heat flux level. The black circles are the experimental data and the red squares are the prediction of the general heat conduction law, they agree well with each other in a wide temperature range.

Based on the above discussions on the heat flux and environmental temperature, the experimental data agree with the theoretical prediction of the general heat conduction law quantitatively. It is the first experimental proof of the thermomass theory.

The uncertainty analysis of the experiment is given as follows:

The thermal conductivity is calculated as $\kappa = \frac{R_0-R_r}{T_0-T_r}\frac{l}{12bw\delta}$, where l, w and δ are the length, width and thickness of the nanofilm. b is the slope of resistance-electrical power data, the best fitted value is obtained using a least square regression method with $R^2 > 0.999$. The uncertainty of κ is given as:

$$\begin{aligned}\ln\kappa &= \ln(R_0 - R_r) + \ln l - \ln(T_0 - T_r) - \ln b - \ln w - \ln\delta - \ln 12\\ &\Rightarrow \frac{\Delta\kappa}{\kappa} \approx \frac{\Delta R}{R} + \frac{\Delta l}{l} - \frac{\Delta T}{T} - \frac{\Delta b}{b} - \frac{\Delta w}{w} - \frac{\Delta\delta}{\delta}\end{aligned} \tag{4.20}$$

where $\Delta T = \frac{R_f - R_0}{R_0 - R_r}(T_0 - T_r)$, thus the uncertainty of ΔT is $\frac{\Delta T}{T} \approx \sqrt{2\left(\frac{\Delta R}{R}\right)^2 + \left(\frac{\Delta T_0}{T_0}\right)^2}$ Finally one can get:

$$\frac{\Delta\kappa}{\kappa} \approx \sqrt{3\left(\frac{\Delta R}{R}\right)^2 + \left(\frac{\Delta l}{l}\right)^2 + \left(\frac{\Delta T_0}{T_0}\right)^2 + \left(\frac{\Delta b}{b}\right)^2 + \left(\frac{\Delta w}{w}\right)^2 + \left(\frac{\Delta \delta}{\delta}\right)^2} \tag{4.21}$$

where $\frac{\Delta R}{R} = \sqrt{\left(\frac{\Delta U}{U}\right)^2 + \left(\frac{\Delta I}{I}\right)^2}$. The uncertainty of each parameter in the Eq. (4.21) is given as:

1. High precise digital voltmeter, $\frac{\Delta U}{U} = \frac{\Delta I}{I} \leq 0.01\,\%$;
2. Temperature controller, $\frac{\Delta T_0}{T_0} \leq 0.1\,\%$;
3. SEM imaging, $\frac{\Delta l}{l} = \frac{\Delta w}{w} \leq 1\,\%$;
4. Calibrated quartz crystal thin-film thickness monitor, $\frac{\Delta\delta}{\delta} \leq 0.1\,\%$;
5. Least square regression method, $\frac{\Delta b}{b} \leq 0.1\,\%$.

Hence, the relative uncertainty of the thermal conductivity is $\frac{\Delta\kappa}{\kappa} \leq 2\%$, the temperature uncertainty is ± 3 K accordingly.

The average temperature of the nanofilm is extracted from the resistance, its relative uncertainty is $\frac{\Delta T}{T} \approx \sqrt{2\left(\frac{\Delta R}{R}\right)^2 + \left(\frac{\Delta T_0}{T_0}\right)^2}$, thus $\left(\frac{\Delta T}{T}\right)_{\max} \leq 1.02 \times 10^{-3} \times 100\,\% = 0.1\,\%$. Assuming $T_{\max} = 300$ K, $\Delta T_{\max} = \pm 0.3$ K.

4.3 Conclusions

1. A DC heating measurement system has been established. The electrical and thermal conductivities of several metallic nanofilms have been measured in a wide temperature range from 2.8 to 300 K. The measurement accuracy has been testified by comparing the measured result of a pure Pt wire with the standard data.
2. The experimental result demonstrates that the elastic Drude's relation is not valid at low temperatures, the electron Raman scattering plays an important role. This leads to the breakdown of MS theory and Wiedemann–Franz law. An effective electron MFP model is developed and matches with the experimental data quantitatively.
3. The steady non-Fourier heat conduction phenomenon has been observed in the Au nanofilm heated by a large current. The measured temperature is remarkably higher than the prediction of Fourier' law, the temperature difference increases as the heat flux increases and the environmental temperature decreases. Meanwhile, the experimental results agree with the prediction of the general heat conduction law quantitatively. This is the first experimental evidence for the validation of the thermomass theory.

References

1. Q.G. Zhang, B.Y. Cao, X. Zhang, M. Fujii, K. Takahashi, Size effects on the thermal conductivity of polycrystalline platinum nanofilms. J. Phys. Condens. Matter **18**, 7937–7950 (2006)
2. K. Fuchs, The conductivity of thin metallic films according to the electron theory of metals. Proc. Cambridge Phil. Soc. **34**, 100–108 (1938)
3. E.H. Sondheimer, The mean free path of electrons in metals. Adv. Phys. **1**, 1–42 (1952)
4. J.W.C. De Vries, Temperature and thickness dependence of the resistivity of thin polycrystalline aluminum, cobalt, nickel, palladium, silver and gold films. Thin Solid Films **167**, 25–32 (1988)
5. A.F. Mayadas, M. Shatzkes, J.F. Janak, Electrical resistivity model for polycrystalline films: the case of specular reflection at external surfaces. Appl. Phys. Lett. **14**(11), 345–347 (1969)
6. A.F. Mayadas, M. Shatzkes, Electrical-resistivity model for polycrystalline films: the case of arbitrary reflection at external surfaces. Phys. Rev. B **1**(4), 1382–1389 (1970)
7. W. Kappus, O. Weis, Radiation temperature and radiation power of thermal phonon radiators using diamond as transmission medium. J. Appl. Phys. **44**(5), 1947–1952 (1973)
8. C.R. Tellier, A.J. Tosser, The temperature coefficient of resistivity of polycrystalline radio frequency sputtered aluminum films. Thin Solid Films **43**, 261–266 (1977)
9. T.Q. Qiu, C.L. Tien, Femtosecond laser heating of multi-layer metalsłi. analysis. Int. J. Heat Mass Transfer **37**(17), 2789–2797 (1994)
10. T.Q. Qiu, C.L. Tien, Femtosecond laser heating of multi-layer metalsłi. experiments. Int. J. Heat Mass Transfer **37**(17), 2799–2808 (1994)
11. A. Vedavarz, K. Mitra, S. Kumar, Hyperbolic temperature profiles for laser surface interactions. J. Appl. Phys. **76**(9), 5014–5021 (1994)
12. J.M. Ziman, *Electrons and phonons, the theory of transport phenomena in solids*, vol. 260 (Clarendon, Oxford, 1960)
13. C.L. Tien, A. Majumdar, F.M. Gerner, *Microscale energy transport*, vol. 28 (Taylor & Francis, Washington DC, 1997)
14. N. Stojanovic, D.H.S. Maithripala, J.M. Berg, M. Holtz, Thermal conductivity in metallic nanostructures at high temperature: electrons, phonons, and the wiedemann franz law. Phys. Rev. B **82**, 075418 (2010)
15. A. Smekal, Zur quantentheorie der dispersion. Naturwissenschaften **11**, 873 (1923)
16. H.A. Kramers, W. Heisenberg, Inverse raman spectra: induced absorption at optical frequencies. Z. Phys. **31**, 681 (1925)
17. E. Schrödinger, An undulatory theory of the mechanics of atoms and molecules. Ann. Phys. **81**, 109 (1926)
18. P.A.M. Dirac, Two-photon processes in complex atoms. Proc. Roy. Soc. Lon. A **114**, 710 (1927)
19. C.V. Raman, K.S. Krishnan, A new type of secondary radiation. Nature **121**, 501 (1928)
20. G. Landsberg, L. Mandelstam, Eine neue erscheinung bei der lichtzerstreuung. Naturwissenschaften **16**, 557 (1928)
21. G. Wiedemann, R. Franz, Ueber die wärme-leitungsfähigkeit der metalle. Ann. Phys. Chem. **89**, 497 (1853)
22. A. Sommerfeld, Zur elektronentheorie der metalle auf grund der fermischen statistic. Naturwissenschaften **15**, 825 (1927)
23. E.W. Fenton, J.S. Rogers, S.B. Woods, Lorenz numbers of pure aluminum, silver, and gold at low temperatures. Can. J. Phys. **41**, 2026 (1963)
24. G.K. White, R.J. Tainsh, Electron scattering in nickel at low temperatures. Phys. Rev. Lett. **19**, 165 (1967)
25. G.S. Kumar, G. Prasad, R.O. Pohl, Review, experimental determinations of the Lorenz number. J. Mater. Sci. **28**, 4261–4272 (1993)
26. C.R. Tellier, L. Ouarbya, A.J. Tosser, Effects of electron scatterings on thermal conductivity of thin metal films. J. Mater. Sci. **16**, 2287 (1981)

27. W.G. Ma, H.D. Wang, X. Zhang, K. Takahashi, Different effects of grain boundary scattering on charge and heat transport in polycrystalline platinum and gold nanofilms. Chin. Phys. B **18**, 2035 (2009)
28. H.D. Wang, J.H. Liu, X. Zhang, Z.Y. Guo, K. Takahashi, Experimental study on the influences of grain boundary scattering on the charge and heat transport in gold and platinum nanofilms. Heat Mass Transfer **47**, 893–898 (2011)

Chapter 5
Conclusions

1. Based on the thermomass theory, the state equation of thermon gas has been established for different materials, such as the ideal gas, dielectrics and metals. Similar to the state equation of ideal gas, the pressure of thermon gas is proportional to the product of its density and temperature. Different micro-distribution functions of the gas moleculars, phonons and electrons will lead to different state equations. At high temperature limit, a unified state equation exists.
2. A two step thermomass model has been established for metals heated by ultra-short pulsed lasers. The governing equation for temperature is a damped hyperbolic equation, which is capable of predicting the propagation of thermal wave. Based on the thermomass theory, the thermal wave is understood as the pressure wave in thermon gas. The numerical simulation results demonstrate that the thermomass theory has clear physical explanation to the thermal wave and causes no negative temperature.
3. A heat flux choking phenomenon has been predicted by the thermomass theory. When the thermal Mach number equals unity, the flow of thermon gas is choked at the end of the nanofilm. A temperature jump occurs at the end with the maximum heat flux. The theoretical prediction agrees well with the experimental results in electrically heated carbon nanotubes.
4. A femtosecond laser transient thermoreflectance (TTR) system has been established. The electron-phonon coupling factor G has been measured for metallic nanofilms. G is an important factor describing the strength of interaction between the electrons and phonons. Simultaneously, the contact thermal resistance R between the nanofilm and substrate can be determined. G is found to be independent of the film thickness, while R increases as the film thickness increases due to the increasing surface roughness and thermal stress.
5. There are two kinds of temperature oscillation exist in metallic nanofilms, i.e. thermal wave and temperature wave. The temperature wave is caused by the periodic temperature boundary conditions within the heat diffusion model. When the laser pulse is 400 fs at room temperature, only the temperature wave can be observed. The measured propagation speed is close to the Fermi speed.

H.-D. Wang, *Theoretical and Experimental Studies on Non-Fourier Heat Conduction Based on Thermomass Theory*, Springer Theses, DOI: 10.1007/978-3-642-53977-0_5,

6. A direct current heating measurement system has been established. The electrical and thermal conductivities of several metallic nanofilms have been measured in a wide temperature range from 2.8 K to 300 K. A spring-pin contact device has been invented to ensure a good electrical connection between the nanofilm and lead wires. All the film samples have been vacuum annealed to remove the defects. The experimental results demonstrate that the elastic MS theory and Wiedemann-Franz law break at very low temperatures, due to the contribution of electron Raman scattering. An effective electron mean free path (MFP) model has been developed and matches well with the experimental data.
7. Under the ultra-high heat flux and low temperature conditions, the measured temperature of the nanofilm is remarkably higher than the prediction of Fourier's law. The temperature difference increases as the heat flux increases and the environmental temperature decreases. The general heat conduction law is able to predict the experimental results quantitatively. This is the first experimental proof for the validation of the thermomass theory.

Zeitfracht Medien GmbH
Ferdinand-Jühlke-Straße 7
99095 Erfurt, Deutschland
produktsicherheit@kolibri360.de